Nature as the laboratory

Nature as the laboratory

Darwinian plant ecology in the German Empire, 1880–1900

Eugene Cittadino

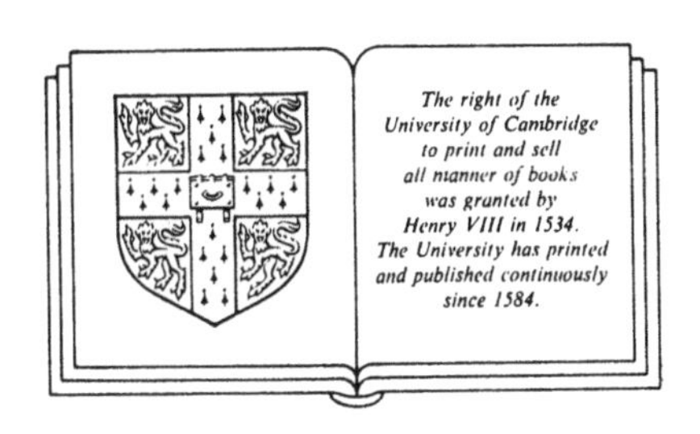

Cambridge University Press

Cambridge

New York   Port Chester   Melbourne   Sydney

PUBLISHED BY THE PRESS SYNDICATE OF THE UNIVERSITY OF CAMBRIDGE
The Pitt Building, Trumpington Street, Cambridge, United Kingdom

CAMBRIDGE UNIVERSITY PRESS
The Edinburgh Building, Cambridge CB2 2RU, UK
40 West 20th Street, New York NY 10011–4211, USA
477 Williamstown Road, Port Melbourne, VIC 3207, Australia
Ruiz de Alarcón 13, 28014 Madrid, Spain
Dock House, The Waterfront, Cape Town 8001, South Africa

http://www.cambridge.org

First published 1990
First paperback edition 2002

A catalogue record for this book is available from the British Library

Library of Congress Cataloguing in Publication data
Cittadino, Eugene.
Nature as the laboratory: Darwinian plant ecology in the German Empire,
1880–1900 / Eugene Cittadino.
p. cm.
Includes bibliographical references.
ISBN 0 521 34045 4
1. Botany – Germany – Ecology – History – 19th century.
2. Botany – Ecology – History – 19th century. 3. Botany –
Germany – History – 19th century. I. Title.
QK900.73.G3C58 1990
581.5′0943′09034–dc20 90-1422 CIP

ISBN 0 521 34045 4 hardback
ISBN 0 521 52486 5 paperback

For my mother and father

CONTENTS

PREFACE

This work is the product of a line of inquiry that began on very general terms and then became more and more refined and focused over the course of several years. As a student of both philosophy and the sciences, I was curious as to the place of the science of ecology in Western culture. Did ecological science emerge out of the same cultural milieu that produced biochemistry and particle physics, or was ecology so fundamentally different from the beginning that its roots lay in an entirely novel set of circumstances and attitudes that have yet to be explored in the history of science? Were ecologists responding to radically different questions than practitioners of other sciences? Did they perceive entirely different problems? Did ecology emerge as a reaction to reductionistic and mechanistic trends in Western science?

This general curiosity led first to some necessary but unproductive speculations and then to a few initial forays into the nineteenth-century background of ecological science. I did not find satisfactory answers to my general questions, but I hope to pursue them in later works. As happens often in historical research, my inquiry took me further and further away from the large questions that led me into this area initially and toward a narrower, but ultimately more manageable, set of secondary issues. Since ecology first appeared as a formal scientific discipline among botanists around the turn of the twentieth century, I focused on a group of American botanists who began to articulate the dimensions of this new field – Conway MacMillan, William F. Ganong, Henry Chandler Cowles, and Frederic Clements, among others. An examination of the intellectual and institutional backgrounds of these early ecologists led me to conclude that the emerging new field was as much a response to professional concerns within botanical science as a response to external issues, such as the recognition of the limits of natural resources and the encroachment of urban industrial civilization on natural environments.

Further inquiry led me to realize that this group of American botanists acknowledged a strong intellectual debt to work that had come out of Germany in the last two decades of the nineteenth century. Casual curiosity regarding these German botanists led to fascination. The Germans had much in common with the Americans – youth, first-rate training in botany, and opportunities to pursue their research outside the laboratory. I discovered not only a group of researchers who had shifted their interests from physiology and morphology to the study of adaptation, but also a group of enthusiasts for Darwin's natural selection theory who had found opportunities to exploit Germany's new interest in colonial expansion in their research. I abandoned my larger search for the roots of ecology and decided to make a thorough study of these German students of plant adaptation – to look into their careers, their institutional affiliations, to place their work against the background of the development of botanical science within the German university system, and to examine in general the motivations for their research. The present work is the result of that effort. I believe that this study does shed light on the general question of the origins of ecological science, as I point out in Chapter 10, but the connection with ecology is only a secondary issue. This is much more a case study of late-nineteenth-century biological science than a treatise on the history of ecology.

There was another unexpected consequence of this undertaking. Although I had prior training in botany, in the course of studying the works of these nineteenth-century German botanists I came to appreciate more fully the intricacies and nuances of plant structure and function. The more I studied, the greater became my fascination not only with the complexities of plant life but also with the complexities of the tasks facing botanical researchers. I hope that the present work conveys something of this fascination to the reader. I hope also that it conveys a sense of the rich source of material that the history of the plant sciences holds for the history of science generally.

Scholarly work tends to be a solitary pursuit, especially in the humanities. This book is no exception. Much of the work was done in isolation, with only a handful of friends and colleagues having an inkling as to what I was up to. Nevertheless, this work could not have been completed without encouragement and support from a number of individuals and institutions. I owe my largest debt to the late William Coleman, under whose direction this project was originally conceived. Bill's combination of tough-mindedness, intellectual integrity, and pure delight in scholarship is all too rare in the academic world. As for many of his students, his presence remains with me as a persistent inner voice that continually asks as I work: "Is that really what you mean to say?" There is always a tone of mild amusement in the voice. I am grateful to the Department of History of Science of the University of Wisconsin and to the university in general for providing a most positive and congenial setting for research and writing – not to mention an outstanding and remarkably accessible library system, supplemented so well by the efforts of history of science bibliographer John Neu. I owe thanks to the Program in Science, Technology, and Society of the Massachusetts Institute of Technology, the

Department of the History of Science of Harvard University, and the Department of History and the Office for the History of Science and Technology of the University of California for providing the facilities where I completed most of the work of revision. The research for this project was aided in part by fellowships from the Exxon Education Foundation and the Research Foundation of the State University of New York and by a grant-in-aid from the American Council of Learned Societies. Special thanks also go to Steve Stephenson and the entire plant ecology group at Michigan State University for taking on a very raw recruit several years ago; to Orie Loucks for helping to sustain my interest in the history of ecology and for giving me ample opportunity to maintain contact with working ecologists; to Malcolm Nicolson, Frank Egerton, Robert McIntosh, and John Beatty for reading the manuscript at one stage or another and offering encouraging words; and to Timothy Allen for several years of good conversation and unsolicited enthusiastic support.

Introduction

As the opening speaker at a special session of the Munich Academy of Sciences in 1898, plant morphologist Karl Goebel chose as his topic "On the Study and Interpretation of Adaptive Phenomena in Plants." He cited three reasons for the recent interest in adaptation among European botanists. To begin with, it seemed to Goebel that the study of the internal structure and development of plants had come to a standstill at the close of the nineteenth century. The really exciting discoveries lay in the past; there was little hope that in the near future researchers would unravel the mysteries of the composition and properties of protoplasm. He likened the situation to that of a mountain climber who views with despair the seemingly insurmountable peaks ahead of him and then suddenly realizes that there is much that is of interest in the maze of ridges and gorges that lie close at hand. Thus perceptive botanical adventurers had recently discovered a host of interesting problems having to do with the relationship of organic structure to the external world, problems that, unlike those dealing with the ultimate nature of the living material, admitted of possible solution. But this seemed a largely negative incentive for studying adaptation. Goebel then suggested two positive sources of stimulation. One was Darwinism; the other was the opening up of the tropics to serious botanical research. With the establishment of modern laboratory facilities, such as that at Buitenzorg, on the island of Java, European botanists could study living tropical plants in their natural setting; they no longer had to rely upon dried specimens or upon observations of plants living in the artificial environment of a greenhouse. The greater taxonomic diversity of tropical vegetation and the wide range of environmental conditions available in the tropics allowed them to investigate adaptive phenomena that were barely discernible in native European plants.[1]

In these brief remarks Goebel accurately characterized the essential features of a movement that had been developing since the early 1880s. Young German botanists trained within a well-established inductive laboratory tradition had ventured out of their laboratories (or rather had transferred their laboratories, figuratively and in some cases literally, outdoors) and had made important contributions to Darwinian biology and to the not yet formalized science of plant ecology. During the early 1880s, the already prolific botanical literature in Germany was marked by an increase in the number of studies dealing with the relationship between the minutiae of internal structures and factors in the external environment of the plant, such as moisture, temperature, and light. This work increased in volume and sophistication during the 1880s and 1890s and rapidly came to include studies of vegetation found in exotic regions. When they gave it a specific name, the Germans usually called this new line of research "plant biology," to distinguish it from purely descriptive botany and from highly specialized, and less interpretive, studies in plant anatomy and physiology. Today this work would fall within the province of plant ecology, specifically autecology, or physiological plant ecology.[2]

Although Darwinism was doubtless a strong factor in the development of research into plant adaptation, Goebel could not have overemphasized the role played by experiences in exotic environments. Opportunities for German botanists to travel in the tropics and in other regions outside Europe were closely linked to imperialism in one form or another. Since the 1860s, the Academy of Sciences of Berlin (officially the Royal Prussian Academy of Sciences) had been sponsoring scientific expeditions to regions where Germans had commercial interests. The expeditions became more frequent after the mid-1880s, when Germany officially acquired "protectorates" in Africa and the East Indies. Ambitious young botanists could convince the Berlin Academy, or a similar regional scientific body, to help finance an excursion to the Egyptian desert or the American tropics, not to mention the new German territories in Africa. In addition, Dutch botanist Melchior Treub established his botanical research laboratory at Buitenzorg just a few years before Bismarck successfully negotiated with the British for control of nearby islands in Micronesia. Thereafter, a steady stream of German botanists paid visits to Treub's laboratory.

Treub would have had few German visitors, however, had it not been for the abundance in Germany of professionally trained botanists with sufficient laboratory experience to make good use of the facilities at Buitenzorg. Botanical programs in German universities had been turning out laboratory-trained botanists at a fairly steady rate since the 1850s. The difficulty of obtaining a permanent university appointment had not discouraged young men from pursuing advanced training in botany. Although established professors as well as *Privatdozents* made the trip to Treub's laboratory, younger botanists who did not have permanent appointments had the time, and the motivation, to make repeated and/or extended visits to the tropics. They could carry out research to help advance their budding academic careers while taking part in the general expansion of German interests beyond the national boundaries of the Reich.

I owe the title of this book to Ernst Stahl, one of these German researchers, who is alleged to have remarked on numerous occasions: "My laboratory is Nature."[3] In many ways, Stahl's background is typical of the botanists considered in this study. Trained by the preeminent plant morphologist and plant physiologist of his day – Anton de Bary and Julius Sachs, respectively – Stahl began his research career making minute investigations into isolated botanical phenomena and then gradually shifted his focus to the relationship between the plant and its environment. The subject of his research became neither plant anatomy nor plant physiology, but plant *adaptation*. Like the others, Stahl felt dissatisfied with the limitations imposed upon botanical inquiry by the narrow inductive laboratory tradition, and he wished to apply his thorough training within that tradition to problems concerning the living plant in its natural habitat.

How and why this interest in plant adaptation should have developed in Germany during the last two decades of the nineteenth century is the central theme of the present work. The approach that I follow is to build up a group portrait of the key figures in this movement, supplying background material where necessary and focusing in detail on some of the central motivations behind and implications of this research program. Although the time constraints are not intended to be strict, the years 1880–1900 represent, for all of the botanists concerned, the period of their most active research regarding plant adaptation. The central figures in this group – Gottlieb Haberlandt, Georg Volkens, Stahl, and A. F. W. Schimper – are relatively unknown even among biologists and historians of biology. The minor figures – Heinrich Schenck, Alexander Tschirch, Emil Heinricher, and Albrecht Zimmermann – are still more obscure. However, all of them were known, and some quite well known, among plant scientists within and outside Germany in the late nineteenth and early twentieth centuries, especially among the growing community of field-oriented botanists who were beginning to articulate the content and methodology of a formal science of plant ecology.

Plant ecology was the first of the various branches of ecological science to be pursued self-consciously, as a discipline in its own right. Animal ecology, limnology, marine ecology, population biology, and other ecological fields developed more or less independently and matured somewhat later than plant ecology. (The first serious glimmerings of a *general* science of ecology appeared only in the 1930s, and the active pursuit of a unified ecological science is largely a post-1960 phenomenon.) Because many of the early plant ecologists directed their attention to the plant community rather than the individual plant, and because the study of plant communities remained a dominant emphasis in plant ecology well into the twentieth century, both ecologists and historians of ecology have searched for the roots of plant ecology in nineteenth-century phytogeography and have ignored, until recently, the environmental emphasis of many of the anatomical/physiological studies in late-nineteenth-century German botany. The attempts by nineteenth-century phytogeographers to relate patterns in vegetation cover to the nature of the physical environment certainly have a central place in the history of plant ecology. However, plant ecology has always endeavored to do more than simply associate vegetation patterns with particular environments; since its beginnings, the science has been preoccupied as well with the *causes* of pattern and change in plant communities. Danish botanist Eugenius Warming, a pioneer community ecologist and one of the principal founders of formal plant ecology, made this point quite clearly in one of the first textbooks in the new field. Warming stated that, in addition to identifying natural plant communities and determining the general physiognomy of vegetation in particular regions, ecology seeks "to investigate the problems concerning the economy of plants, the demands that they make on their environment, and the means that they employ to utilize the surrounding conditions and to adapt their external and internal structure and general form for that purpose."[4] As such, plant ecology has had to rely on the kinds of detailed, physiologically oriented studies relating habitat factors to plant structure that characterize the work of the German botanists under consideration here. Despite the descriptive nature of many of the early formal contributions to plant ecology, both European and American plant ecologists recognized a debt to these German ecophysiological botanists and continued to insist on the close connection between plant ecology and physiology.

For their part, however, the Germans who initiated plant adaptation studies in the late nineteenth century were not seeking to found a new scientific discipline. They were seeking only to broaden the scope of botanical science by redirecting the focus to the plant in its natural surroundings. Their botanical training had a strong physiological orientation and, equally important, they were among the first generation of German botanists to come of age, so to speak, within a Darwinian universe; they attended universities in the late 1870s and early 1880s, when Darwinism was enjoying its greatest popularity in Germany. With their backgrounds in plant anatomy and physiology, they saw in the concept of natural selection the key to explaining the manifold complex adaptations of plants to biotic and abiotic factors in their environments. Their Darwinism convinced them that every aspect of plant structure has a function in the living plant, and they set about developing programs to exploit the research potential of that principle. As much as any single group of biologists in the nineteenth century, this group of German botanists attempted to apply the concept of natural selection directly to their individual research projects. This was a relatively rare phenomenon, since the principal effect of Darwin's evolution theory had been to direct the attention of biologists to problems relating to *phylogeny*. In 1927, for example, British ecologist Charles Elton complained that Darwin's evolution theory had sent zoologists flocking indoors for over fifty years (to search for phylogenetic relationships in comparative studies of external and internal animal structures) and that only recently had some of them begun to venture out into the open air to study animals in their natural surroundings.[5] The situation in botany was not nearly so bad, but it was certainly the case that few botanists had used the concept of natural selection as a justification for studying adaptation. Stahl, Haberlandt, Schimper, and their colleagues represent notable exceptions.

If Darwinism was an important stimulus for their ecological research, so also was their field experience in foreign lands. As a result of Germany's colonial expansion in the last two decades of the nineteenth century, opportunities developed for German scientists to carry on their research far beyond the political boundaries of the Reich. The botanists under consideration in this study were especially well prepared to take advantage of these opportunities, since they had a research program designed to investigate the relationship of environmental factors to organic form and function. Particular adaptations are often easier to recognize in organisms living under unusual or extreme environmental conditions; and most of the travel opportunities were in the tropics, where plants exist under conditions of heat and moisture quite unfamiliar to botanists trained in central Europe. Firsthand experience of the diverse array of adaptive phenomena exhibited by tropical plants, therefore, served to reinforce the environmental emphasis of their work. Their most important ecological studies involved non-European plants. Nevertheless, they continually stressed the need to apply the knowledge of plant adaptation gained in exotic lands to an understanding of adaptive phenomena in native European vegetation. They did not have practical applications in mind, at least not at first; they simply wanted to use their experience outside Europe to gain insights into the organization and behavior of the familiar plants upon which most of their anatomical and physiological knowledge was based.

This early Darwinian school of plant ecology was very much an outgrowth of the cultural and institutional settings in which it developed. Chapter 1 examines the major institutional and intellectual themes that served as the background for botanical work in Germany in the late nineteenth century. Important here was the professional development of botanical science during the third quarter of the nineteenth century and its relationship to the German university system. The researchers who turned their attention to studies of plant adaptation in the 1880s were all trained within a thorough laboratory tradition that had become the mark of German botany. This tradition was born of several causes – the reaction to speculative science, the improvement in microscopic technique, the development of research seminars and institutes within the German universities – and it was characterized by a cautious, inductive, compartmentalized approach to botanical problems. Once they received their training within this tradition, botanists could anticipate a lengthy period of internship before finding a permanent university appointment, since the number of professorships in botany, as in most fields, had not kept pace with the growing ranks of postdoctoral students. This long period without the security or the responsibilities of a permanent appointment may, in some cases, have encouraged research in nontraditional areas and contributed to extended botanical excursions in exotic regions.

In terms of influence and intellectual traditions, the German eco-physiological botanists fall into two broad groups. The first group consists of students of Simon Schwendener, who explored an approach to the study of plant life that they chose to call "physiological plant anatomy." This group was centered at Schwendener's botanical institute at the University of Berlin, although physiological plant anatomy itself was first developed by Gottlieb Haberlandt, who studied with Schwendener for a short period in Tübingen. Chapter 2 examines the careers of Schwendener and Haberlandt, their brief collaboration, and some of the difficulties that their research program encountered with proponents of the traditional compartmentalized approach to botanical research, which maintained the separation of anatomical and physiological investigations. Haberlandt, more so than Schwendener, was criticized for mixing together discussions of structure and function, a practice that he justified on the basis of Darwin's natural selection theory. Chapter 3 treats the Darwinian character of Haberlandt's work, and Chapter 4 surveys briefly the work and the personnel of the Schwendener school in Berlin. A number of Schwendener's students applied this new approach to botany outside the laboratory and outside the boundaries of Europe. Chapter 5 examines the research carried out by Georg Volkens in Africa and by Haberlandt in Java and Malaysia and briefly discusses the background of German colonial development.

The second group of plant adaptationists consists of students trained at Strasbourg and Bonn by Anton de Bary and Eduard Strasburger, neither of whom had a strong personal interest in the problem of adaptation. Two members of this group, A. F. W. Schimper and Heinrich Schenck, were longtime colleagues at Bonn who carried out extensive field research in South America. The third member, Ernst Stahl, had the good fortune to secure a permanent position relatively early at the University of Jena, where he conducted his research somewhat in isolation from the others. Chapter 6 follows Stahl's research from his early straightforward anatomical and physiological investigations under the direction of de Bary and Sachs to his elaborate studies of adaptive phenomena in Europe and abroad. Chapter 7 traces the careers of Schimper and Schenck, focusing on their research in the American tropics. These three together produced perhaps the most extensive and detailed Darwinian treatments of plant adaptation to be found among botanists in the late nineteenth century.

Chapters 8 and 9 concern the primary sources of motivation behind this ecological work: Darwinism and field experience in exotic lands. Chapter 8 examines particular examples of the use of the concept of natural selection to interpret adaptive phenomena, the relationship between this approach to adaptation and that employed by German botanists earlier in the century, and the criticisms raised by contemporaries regarding this Darwinian school of plant biology. Those criticisms, for the most part, were directed at the teleological character of the plant adaptation studies, but the authors of these studies were convinced that Darwin's evolution theory justified the return to discussions of purpose in biological science. Chapter 9 considers the connection between these studies of plant adaptation and German colonial expansion. These botanists did not generally take an active interest in the economic development of the colonies during the period to which this study is confined. Instead, they took advantage of Germany's imperial venture to extend their research experience – to engage, as it were, in a kind of intellectual imperialism. However, their endeavors were aided considerably by scientific facilities, such as the Buitenzorg Botanical Garden, which were tied closely to the economic interests of the colonies.

In Chapter 10 I briefly point out some of the specific connections between the late-nineteenth-century German studies in plant adaptation and the early formal development of plant ecology, and I offer speculations regarding the broader relationship between these two movements. The central point here is that the appearance of a formal science of plant ecology in the early twentieth century can be viewed as the result of the integration, under favorable institutional settings, of the causal/functional approach to botanical science (as characterized by the work of these Germans) with the phytogeographical tradition of identifying and classifying natural plant communities. Although the United States served as the most active center of the new discipline, American plant ecologists, as well as those in England and Scandinavia, were influenced, directly or indirectly, by the ecophysiological studies that came out of Germany in the late nineteenth century.

Contemporary critics of neo-Darwinism, such as Stephen Gould, would characterize this group of German botanists as adaptationists; they strongly believed that their adherence to a strict selectionist view of organic transformation compelled them to regard virtually all plant structures as having adaptive significance.[6] As such, they differed from most of the early plant ecologists, whose positions regarding the mechanism of evolution varied between neutrality and Lamarckism. However, although the science of plant ecology may have drawn its formal lines during a period when Darwinism was out of favor, the practitioners of the new science were quite cognizant of the force of the Darwinian explanation for evolution and the attention it directed to the problem of adaptation. Although the focus of the new plant ecology was on the nature, rather than the origins, of adaptations, early plant ecologists made use of, and paid tribute to, the pioneering work of these late-nineteenth-century Germans, for whom Darwinism had been a primary factor in redirecting their intensive botanical training from the laboratory to the natural habitat of the plant. Yet whatever their influence on twentieth-century ecological science, studies of the relationship between the living plant and its environment came about in Germany during the last two decades of the nineteenth century as the result of a combination of intellectual trends, professional concerns, and political developments peculiar to that time and place. This is as much a study of the history of German science, Darwinian biology in the late nineteenth century, the institutionalization of the life sciences in Germany, and the relationship between science and colonialism as it is a study of the early history of plant ecology. It is a profile of a group of similarly trained young scientists responding to a common set of intellectual influences and institutional pressures and taking advantage of particular political circumstances. The result of these influences and responses was the appearance, at the end of the nineteenth century, of a Darwinian school of botanical science that never found a permanent institutional home but that nevertheless influenced the development of plant biology and the new science of plant ecology well into the twentieth century.

1

Botany in Germany, 1850–1880: the making of a science and a profession

The interest in plant adaptation that surfaced in the last two decades of the nineteenth century represented a marked change of emphasis in German botany. Descriptive microscopic studies, narrowly defined laboratory investigations, and a very cautious attitude toward generalization characterized the period roughly from 1850 to 1880. This predominantly inductive approach to the study of plants was the result of a reaction to the speculative mood in the natural sciences in Germany during the first third of the nineteenth century, but it also received considerable impetus from the new discoveries made possible by technical advances in microscopy and by the development of laboratory techniques for plant physiology. With only slight exaggeration (and perhaps with a little regret),[1] one could say that in the mid-nineteenth century, botany in Germany was transformed from a branch of natural history practiced by a few university professors, physicians, and amateurs to a laboratory science practiced by a growing cadre of professionals.

By the 1870s and 1880s the expansion of the German university system and its growing emphasis on the sciences led to a steady increase in the numbers of assistants, *Privatdozents*, and assistant professors associated with botany departments, although it did not lead to a dramatic increase in professorships. Overall university enrollment in Germany roughly doubled during the period from 1850 to 1880, and enrollments in philosophical faculties, where botany chairs were located, increased threefold. Some universities created new chairs to accommodate these increased enrollments, but growth in teaching staffs came mainly as an increase in the number of instructors below the level of professor. At the same time, science faculties developed ancillary facilities for teaching and research, thus expanding the number of lower-level positions. In botany, as in other scientific disciplines, this development usually took the form of a special laboratory or institute, with its small staff of assistants under the professor.

Botanists in these new facilities mined with considerable profit the research territories opened to them by the achromatic compound microscopes and by the application of anatomical and physiological laboratory techniques to the study of plants. Although the taxonomy of the seed plants was still the major preoccupation of botanists in other countries, German botanists in the 1850–80 period expanded the scope of their work to include the simpler vascular plants, the bryophytes, the algae, and the fungi, and shifted their focus from taxonomy to anatomy and physiology. Although much of their work was primarily descriptive in nature, and although they maintained an outward vigilance against speculative science, German professional botanists during this period were nevertheless guided in their research by certain broad theoretical considerations and methodological principles: the cell theory, the concept of the alternation of generations in plants, the theory of evolution, and a preference for mechanistic explanations of biological phenomena. The careful pinning of numerous fragments of information onto this general theoretical framework during the 1850s, 1860s, and 1870s provided the background for the more synthetic work of a later generation, which was to direct the accumulated botanical wisdom of its mentors to the problem of the adaptation of plants to their natural environmental conditions.

ESTABLISHING A TRADITION: SCHLEIDEN AND HOFMEISTER

The inductive approach that dominated German botany for most of the nineteenth century began to take hold during the 1840s and 1850s, and it involved a shift in emphasis from the earlier preoccupation with taxonomy and gross morphology. In contrast to that earlier emphasis on the comparative study of fully developed plant forms (the dual legacy of Linnaeus and Goethe), the new botany emphasized the study of the developmental stages, minute anatomical structures, and internal dynamics of individual plants.[2] Although no single discovery or single work effected this change, two books published almost a decade apart stand out as the works that best exemplified the changes taking place within botanical science and set the tone for research and writing for several decades. The first of these works, Matthias Jacob Schleiden's *Gründzuge der wissenschaftlichen Botanik*, first published in 1842–43, was as much a call to action and the sketch of an outline for future research as it was an original contribution to botanical knowledge. The clearest and most influential manifestation of Schleiden's program for scientific botany came in the second work, Wilhelm Hofmeister's *Vergleichende Untersuchungen* of 1851. Taken together, these two works defined the major problems and set the standards for a generation of German botanists.[3]

Schleiden's work was a direct attack on speculative botany and speculative science in general. He began the preface with the sentence "Whoever expects to learn botany from the present book may just as well set it aside unread, since botany is not learned from books."[4] The reader who went ahead despite this warning may well have found justification in Schleiden's remark, since the book contained erroneous statements and misinterpretations, even within the context of botanical knowledge during Schleiden's time, with regard to cell formation and the nature of fertilization in plants. However, Schleiden claimed only to be providing a guide whereby botanists could avoid pitfalls and blind alleys. The first section of the book, a 166-page methodological introduction, offered an elaborate renunciation of philosophical speculation and an outline for the development of botany as an inductive science. Schleiden directly attacked the scientific followers of Fichte, Schelling, and Hegel for mistakenly resting their work on ideas and their interrelationships rather than on observation and experiment. Whatever these *Naturphilosophen* have accomplished, he remarked, they owe not to their philosophy but to the stubborn instincts of genius.[5] Yet despite the rather strong polemics, Schleiden's call to action was not the thoroughgoing materialistic, reductionist program that would appear a few years later among Berlin-based physiologists, such as Hermann von Helmholtz and Emil Du Bois-Reymond. It represented instead a combination of commitment to inductive science and respect for the integrity of the organism grounded in a Kantian approach to the life sciences. Schleiden stated repeatedly that one must study plants, not books; that the object of botanical science is the whole living plant, not the particular parts; and that one cannot expect botany to follow the same laws and principles as physics and chemistry.[6]

Schleiden was attacking what he regarded as the empty speculative trend in early-nineteenth-century science, a trend that expressed itself among botanists, for example, in attempts to explain affinities between plants, and relationships between plant parts, as expressions of regular mathematical or geometric laws. For example, Karl Schimper and Alexander Braun, both of whom had been influenced strongly by Oken and Schelling, attempted to explain the ordering and positioning of leaves on a stem as an expression of geometric rules corresponding to the construction of spiral configurations. Similarly, Alexander von Humboldt associated the number of plant genera and species present in particular regions of the world with recurrent patterns of fixed ratios.[7] For Schleiden, the problem with this approach was that botanists tended to identify the mathematical laws as the causes of the observed regularities in nature and did not seek the causes in the natural phenomena themselves. There was no reason to assume that a search for mathematical regularities would be as productive in botany as it had been in physics. A scientific botany must rest firmly on its own foundations.

Schleiden, for his part, was quite clear regarding the direction that future botanical research should take:

> My aim is to establish the necessity of embracing, as a fundamental principle in the study of the whole, the existence of an essential life in each separate cell. Hence arises the necessity of carrying on investigations in the first

> instance in the individual cells, or in portions of the vegetative structures in which we have to do with few cells in combination. On these we must make our first experiments, and from them draw our first conclusions, which we may then proceed to apply to subsequent investigations into the general structure of the plant, pursuing all our inquiries with the aid of the microscope, and placing them under the control of an accurate history of development. Upon such a plan alone can we make a sure advance in the study of vegetable life.[8]

Thus Schleiden grounded the idea of the integrity of the living organism in the individual living cell. One need reduce the problem no further: One begins with the living cell and then builds up the investigation to include ever higher degrees of complexity. In effect, Schleiden's view represented a compromise between the vitalism of the *Naturphilosophen* and the more extreme forms of mechanistic materialism that would reduce all living phenomena to chemistry and physics.[9] Due partly to new trends already established and partly to Schleiden's clear formulation of these fundamental methodological principles in his textbook, cytology, embryology, developmental history, and physiology became the central preoccupations of a growing number of German botanists during the 1850s, 1860s, and 1870s.

A plan, however, is not enough. In order to sustain itself, a new scientific program should lead to new insights, to new discoveries and generalizations. Here lies the significance of Hofmeister's work. Influenced by Schleiden's textbook, Wilhelm Hofmeister, a Leipzig bookseller and self-taught botanist, made exacting microscopic investigations into the life histories of various plants – mosses, liverworts, lycopods, horsetails, ferns, and conifers – and discovered a fundamental pattern, the so-called alternation of generations, that links all of these plants as particular manifestations of one basic plan. In striking contrast to the bold and polemical style of Schleiden, Hofmeister's style was unpretentious and cautious. Dispensing with all introductory remarks, he opened his *Vergleichende Untersuchungen* as follows, with a description of a familiar bryophyte:

> The germinating plant of *Anthoceros* appears as a circular, more or less lobed, extension of juicy, dark-green cell tissue. Leaves are lacking entirely; the underside of the leaflike flat stem is attached to the solid by means of numerous root hairs. The branching is on the whole bifurcate, more clearly so in *A. laevis* than in *A. punctatus*.[10]

Thus began the work recognized by the next generation of German botanists as the single contribution most responsible for bringing botany out of its middle ages and into the modern period. Hofmeister continued for the next 138 pages with straightforward descriptive prose, as in the preceding passage, and then offered a 3-page conclusion that introduced a fundamental new concept into the botanical literature. He discovered that all green land plants have life cycles that are almost identical in detail, represented by the regular alternation of two generations. In the first generation, a plant, or two separate plants, forms structures that produce two types of sexual cells, or gametes. The gametes then combine in the larger of these structures to form the mother cell of the second generation. The second generation, whether it be the tiny fruiting structure of a moss or a full-grown pine tree, produces spores that germinate to form the first generation, and the cycle begins anew.[11]

This simple concept formed the fundamental unifying principle of German plant science in the mid-nineteenth century. What was so remarkable about Hofmeister's observations, which were repeated in laboratories and classrooms throughout Germany during the next decade, was that particular features of the two generations, such as the form of the male and female reproductive structures, are strikingly similar in detail throughout the plant kingdom. These similarities, which are easily observable under a microscope, provided a concrete material basis for unity in the plant kingdom, a unity grounded neither in sweeping vitalistic doctrines nor in contrived mathematical formulae, but in simple, repeatable observations or regular sequences in cellular differentiation. To many botanists, this discovery came as a revelation. Julius Sachs was effusive in his praise for Hofmeister: "The results of the investigations published in the *Vergleichende Untersuchungen* in 1849 and 1851 were magnificent beyond all that has been achieved before or since in the domain of descriptive botany."[12] The immediate effect of Hofmeister's work was to bring attention to the "lower" forms of plants, the so-called cryptogams (plants that do not form seeds, such as mosses, horsetails, lycopods, and most ferns) and to stimulate investigations into developmental history. Within the next thirty years virtually all of the life cycles of plants, including the algae and most fungi, were worked out in detail by patient researchers who, like Hofmeister, took care to report their observations exactly as they appeared and in painstaking detail. Although few expositions were quite as dry as *Vergleichende Untersuchungen*, unfortunately the language of Hofmeister's work, as well as his approach, became a standard for botanical writing. Younger botanists perhaps took Schleiden's polemic against speculative science too much to heart and emulated Hofmeister's style as well as his methods.

Although the work of Schleiden and Hofmeister exerted its greatest influence on the development of German botany in the middle years of the century, this work did not occur in isolation. Carl Nägeli, for example, a Swiss-born botanist and student of Schleiden, published a study of cell growth in mosses and liverworts in which he traced cell divisions back to a single apical cell.[13] This 1845 work formed the basis for the fundamental botanical concept, worked out later in detail by Nägeli, that, with a few notable exceptions, new cells arise in plants in only one of two ways: by division of an apical cell at the growing tip (meristem) of a root or shoot or by division of a cell in the cambial region within the stem. During the same year in which Hofmeister published his seminal work, Hugo von Mohl published perhaps the clearest exposition to date regarding the state of cellular anatomy and physiology.[14] Von Mohl offered a detailed account of plant cell division and, like Nägeli, expressed his skepticism of, but did not discount, Schleiden's theory of free-cell formation (the notion that cells can arise *de novo* from the proper medium). As regards physiology, von Mohl discussed such processes as the absorption of water, diffusion of sap, respiration, and so on as distinctly cellular processes. Yet, recognizing the still rudimentary state of plant physiology in the early 1850s, he wrote:

> Unhappily, the Physiology of Plants is a science which yet lies in its earliest infancy. Few of its dogmas can be regarded as settled beyond doubt; at every step we meet with imperfect observations, and consequently with the most contradictory views; thus, for example, opinions are still quite divided regarding the doctrines of the development of the cell, of the origin of the embryo, and of the existence of an impregnation in the higher Cryptogams.[15]

To the credit of German botany, all of these matters were settled with reasonable satisfaction within the next twenty years by younger men, such as Nägeli, Nathaniel Pringsheim, and Anton de Bary. In addition, by the 1870s, largely through the efforts of Julius Sachs, the study of plant physiology became a science on a par with animal physiology.

MICROSCOPES AND INSTITUTES

The work of Schleiden, Hofmeister, Nägeli, von Mohl could not have been possible without the microscope; and the rapid dissemination of their ideas depended upon the general availability of accurate microscopes with which other botanists could repeat their observations. Schleiden stated in his textbook:

> Considering the experience of the last thirty years, there is no need to observe that a profound study of any of the natural sciences, and especially of organization, is impossible without the aid of the microscope. He who expects to become a botanist or a zoologist without using the microscope is, to say the least of him, as great a fool as he who wishes to study the heavens without a telescope.[16]

Microscopes are relatively delicate instruments that are best kept in a room where they can be left undisturbed. They also require space in which to store, preserve, and prepare specimens for observation. Due to these modest requirements, these instruments played a role in the reorganization of botanical instruction and in the development of the botanical laboratory, or "institute," as the Germans preferred to call it.

Although there were some improvements in the quality of lenses around the turn of the nineteenth century, compound microscopes, using lenses with different refractive characteristics to eliminate most of the chromatic aberration, were first manufactured in Europe in the 1820s and became generally available in Germany in the early 1830s. Most of the achromatic microscopes in use in Germany at that time were produced either by the firm of Plössl in Vienna or by that of Pistor and Schiek in Berlin. The advantage of the new instruments was not so much in the strength of magnification as in the clarity of the image, or at least in the perceived clarity of the image.[17] Jan Evangelista Purkyně, a physiologist at the University of Breslau, expressed his enthusiasm over the new instruments in the following manner:

> With the acquisition of the Plössl microscope in the summer of 1832, a new epoch in my physiological activity began. Everyone who has made a serious trial of the new instrument knows that our vision is so augmented by it that all limitations of normal vision are removed and new worlds are being discovered on all sides. With inevitable wolf-hunger, I have investigated all types of tissues, animal and plant, and became convinced of the inexhaustibility of new possibilities.[18]

Purkyně became an important vector for change. He had already set up a physiological laboratory in his home in 1824, a year before Justus Liebig established his chemical laboratory in Giessen. Through continued efforts, Purkyně eventually persuaded the Breslau authorities to establish a physiological institute, the first of its kind in Germany, in 1839.[19]

The examples set by Liebig and Purkyně established the pattern for scientific research institutes associated with the German universities. Chemistry and physiology led the way; botany and zoology lagged somewhat behind. Botanist Eduard Strasburger credited Anton de Bary with setting up the first botanical institute in Germany, at the University of Freiburg-im-Breisgau, in 1858. This was a one-room affair without separate funding from the university. In the mid-1860s de Bary established a somewhat larger laboratory at Halle. By that time, botanical laboratories had been built in Breslau, Munich, and Jena, and the new structures came with separate endowments.[20] This was all part of the general trend toward the creation of specialized research seminars and institutes within the German universities in the middle third of the nineteenth century. The bulk of the budget increases in Prussian universities between 1820 and 1870 went toward the establishment and maintenance of seminars and institutes.[21] De Bary's institute at Halle became one of the most active centers for botanical research during the 1860s, and those of Nathaniel Pringsheim at Jena and Carl Nägeli at Munich also attracted a number of young botanists. In 1872 de Bary moved from Halle to the reorganized University of Strasbourg, where he set up a model botanical institute that attracted students from all over the world. By that time Julius Sachs, who had served as Purkyně's laboratory assistant at Prague in the 1850s, established at the University of Würzburg the only botanical institute that would rival de Bary's in popularity. Under Sachs the Würzburg institute became the international center for plant physiology research from the late 1860s through the 1880s.[22]

The equipment within these botanical laboratories was relatively unsophisticated. The microscopes were variations on the compound achromatic instruments developed in the 1830s. Although microscopic technique improved steadily during the century, the next significant change did not occur until the late 1870s, with the introduction of oil immersion lenses, better staining techniques, and improved mechanical devices for cutting thin sections.[23] The botanical work done before 1880 or so did not benefit from these later changes, but the instruments and techniques available to botanists in the middle years of the century were adequate for their purposes. With the compound achromatic microscopes, researchers were able to make accurate observations of stems and leaves in cross section, germinating spores and developing embryos, algal cells, and fungal cells. Along with the microscopes, the laboratories were equipped with a few reagents and fixatives, for preserving and staining specimens, and with simple apparatus for physiological investigations. During the 1860s and 1870s, Julius Sachs gradually introduced more sophisticated equipment and techniques for growing plants in the laboratory, and invented a host of ingenious devices for measuring the movements of plants in response to light, gravity, temperature, and other external factors.[24]

The "institutes" varied considerably in size, from one or two rooms, such as those at Tübingen and Halle, to separate structures with several rooms, such as those at Würzburg, Strasbourg, and Jena. The larger facilities had a separate lecture room, a classroom for undergraduates, a laboratory for advanced students, an office and a separate laboratory for the professor, and space for an herbarium and a small library. Beginning students heard lectures, saw demonstrations, and performed a minimum of laboratory work. Advanced students had their own space in the laboratory, where they were expected to work eight to ten hours a day. In addition to the professor, there were one or two assistants and, depending upon the size of the institution, one or two assistant professors and a varying number of *Privatdozents*, all of whom also worked regularly at the institute.[25] Within such physical settings, the program outlined by Schleiden in the 1840s took form and bore fruit. Although German botanists did not neglect taxonomy and collecting, their work turned more and more to the kinds of studies that lent themselves best to laboratory investigation. Apart from the excitement of the new discoveries themselves, the existence of these special institutes – along with their hierarchical and compartmentalized organization (i.e., the professor overseeing the activities of advanced students, instructors, and assistants, each of whom had his assigned space within the laboratory) – served to reinforce the already growing interest in anatomical and physiological investigation.[26]

THE NEW AGENDA

Working with their microscopes and at their laboratory benches in the new institutes, German botanists applied themselves to the task of determining the exact nature of the life cycles, anatomical structures, and physiological processes of plants. Carl Nägeli continued his work in cytology and elaborated his investigations into apical and cambial growth; Nathaniel Pringsheim examined the nature of sexual reproduction in algae, and thus awakened interest in the microscopic investigations of that group of plants; numerous botanists joined Wilhelm Hofmeister in investigating the life cycles of the cryptogams; Anton de Bary initiated the study of the developmental history of fungi and the many forms of fungal parasitism on the higher plants; August Eichler, Hermann Vöchting, and many others initiated detailed studies of the morphology of the flowering plants; and Julius Sachs almost single-handedly transformed plant physiology into a modern experimental science.

If the careful recording of observations characterized this work, there were, nevertheless, certain guiding principles that determined the direction of research and gave form to the interpretation of results. The cell theory clearly played a dominant role; the focus of German botanical research from the 1850s on shifted from the gross morphology of plants to individual cells and tissues. The alternation of generations formed a second guiding principle closely tied to the first. Hofmeister's work inspired a flurry of activity directed at comparing the life cycles of plants at the cellular level; spores, embryos, and reproductive structures received considerable attention. There was a decided preference for mechanistic explanation, a deliberate effort to avoid the errors of the earlier speculative school. Botanists took pains to demonstrate that plant reproductive structures are the results of sequences of cell division and differentiation clearly observable in intricate detail; complex processes, such as seed and spore production, fertilization, and flower formation follow strict causal sequences and obey mechanical principles. Finally, evolution theory provided the broad framework for discussing interrelationships between plants. Botanists immediately saw the connection between Darwin's theory and Hofmeister's alternation of generations: Plants are related to each other by descent from common ancestors; hence the life cycle that Hofmeister found ubiquitous in the land plants was probably the pattern established by the algal form that first successfully colonized the land and gave rise to the more complex forms.[27] Although there was little room for the concept of natural selection in the day-to-day laboratory work of that period, the theory of descent nevertheless provided a general plan of organization that lay in the background at all times.

Schleiden's *Gründzuge der wissenschaftlichen Botanik,* which went through four editions, remained the central botanical textbook until the

1860s.[28] However, a number of excellent smaller textbooks covering special aspects of botany in more depth than that of Schleiden, appeared throughout the 1850s and 1860s. Hugo von Mohl's *Gründzuge der Anatomie und Physiologie der vegetabilischen Zelle,* mentioned earlier, belongs to this category, as does a similar work by Nathaniel Pringsheim published in 1854.[29] In the 1860s, at the suggestion of Julius Sachs, Hofmeister decided to edit a series of volumes under the general heading *Handbuch der Physiologischen Botanik,* whereby "physiological botany," following the custom of that time, referred to anything not covered by taxonomy. This series included works by Hofmeister on cell theory and general plant morphology; Julius Sachs's *Handbuch der Experimental-Physiologie der Pflanzen,* the first complete plant physiology textbook in modern terms; and Anton de Bary's *Morphologie und Physiologie der Pilze, Flechten und Myoxomyceten,* one of the first comprehensive treatments of the fungi, lichens, and slime molds.[30] Collectively, these works represented the state of botanical knowledge in the 1860s. However, more than twenty years had lapsed since the publication of Schleiden's textbook and no one had attempted to gather all of the separate strands of the new research together under one cover. Sachs accomplished this in 1868 with his *Lehrbuch der Botanik.* This textbook went through four editions in only six years and quickly became the standard work in general botany, not only in Germany but in the rest of Europe and in the United States as well.[31]

SACHS'S *LEHRBUCH*

The central figure in German botany from the mid-1860s to the mid-1880s was clearly Julius Sachs (1832–97). A native of Breslau, Sachs was forced to discontinue his secondary education after both of his parents died in the late 1840s. Jan Purkyně, who had been a friend of Sachs's family, offered to take the artistically inclined Julius on as his personal assistant, mainly as an illustrator, at the University of Prague in 1851. Thus employed, Sachs was able to complete his *Gymnasium* education and enter the university, from which he received his doctorate in 1856. After brief teaching experiences in agricultural and technical schools, he was eventually appointed to a chair in botany and natural history at the Agricultural Academy at Poppelsdorf, near Bonn. In 1867, after six years at Poppelsdorf, he was called to a university chair at Freiburg-im-Briesgau, and in the following year he was appointed to the chair in botany at the University of Würzburg, where he remained for the rest of his life.[32]

Through the work of his institute at Würzburg, through his very popular lectures, and especially through his textbook, Sachs achieved for botany the unity and scientific respectability that Schleiden had only dreamed of. Both inside and outside Germany, Sachs's *Lehrbuch der Botanik* generated an excitement that botany has seldom experienced. F. O. Bower wrote that

Sachs's textbook "came as a revelation" to a small group of botanists at Cambridge University in the 1870s: "We felt then that we were daily seeing things not, it was true, new to science, but at least observed for the first in Britain."[33] Sachs's was the first textbook to introduce the cytological discoveries of Nägeli, Hofmeister, and their contemporaries to a general scientific audience. This work was indeed revelatory in the English-speaking world in the 1870s. A deliberate adaptation of Sachs's textbook, Charles Bessey's *Botany for High Schools and Colleges,* became the standard textbook in the United States through the end of the nineteenth century.[34] Sachs's skill as a draftsman played no small part in the success of his textbook. His excellent detailed illustrations continued to be utilized in botanical textbooks for decades, sometimes apparently without acknowledgment to Sachs.[35]

A review of the contents of the fourth (1874) edition of the *Lehrbuch* provides an excellent overview of the major concerns of German plant science in the 1870s. Sachs divided the textbook into three parts – general morphology, special morphology, and physiology – reflecting the standard botanical curriculum in the German universities. During each academic year the professor at a given institution generally alternated his lecture sequence, devoting one semester to general botany (which included morphology and sometimes physiology) and the other semester to special morphology (which involved a comprehensive survey of the entire plant kingdom by taxonomic groups). Although Sachs called the first section of his textbook "General Morphology," it treated most of the topics that today fall under the heading of plant anatomy. The second section, "Special Morphology," dealt with the developmental histories of the major plant groups. In the section on physiology, Sachs began with a discussion of molecular forces and chemical reactions in plants and went on to treat various vital phenomena, such as growth, irritability, and reproduction. Sachs's own physiological research focused on photosynthesis and on the influence of various external factors on plant growth and development. He incorporated the results of this work into the physiological section of his textbook.

In the twenty-five years since the appearance of the first edition of Schleiden's textbook, the inductive program had been so effective that most of the botanical literature covered topics that were too specialized to appeal to the beginning student. Sachs attempted to restore to general botany a sense of the plant as a living organism. This he accomplished quite effectively, largely by introducing students to the methods, tools, and results of physiological investigation. A dedicated academic scientist exposed to both the recent successes in medical physiology, through his association with Purkyně, and to the practical side of plant science, through his years of working in agricultural institutions, Sachs brought to the study of plant physiology a unique blend of hard-headed empiricism, familiarity with experimental technique, and commitment to high scholarly ideals.[36] The section on physiology in his general

Julius Sachs. Courtesy of the Hunt Institute for Botanical Documentation, Carnegie Mellon University, Pittsburgh, PA. Reproduced by permission of G. Fischer, Stuttgart.

botany textbook was larger than his entire *Handbuch der Experimental-Physiologie der Pflanzen,* which had appeared a few years earlier, and it included illustrations and descriptions of many of the special pieces of apparatus that he had designed for the plant physiology laboratory. From the mid-1860s on, largely through Sachs's efforts, physiology received progressively more attention among German botanists, and a physiological point of view found its way into anatomy and morphology as well. For many young botanists, Sachs's textbook provided their first introduction to the subject. Many of them came to Würzburg to hear his lectures, which were as impressive as his textbook, and to work in his laboratory.[37] The best of his students went on to set the standards for research in the next generation: Hugo de Vries made numerous contributions to plant physiology, in addition to his better-known work in hereditary theory; Jacques Loeb greatly extended Sachs's work on tropisms; Wilhelm Pfeffer established an international center for the study of plant physiology at Leipzig; and Karl Goebel established the science of experimental plant morphology by adapting physiological methods to the study of plant development.[38]

If Sachs's textbook brought attention to physiology, it also provided one of the few clear accounts by a botanist of Darwin's evolution theory. Carl Nägeli had published a general treatise on evolution in 1865, criticizing Darwin for his overemphasis on natural selection; but one can find little else of significance in the literature besides Eduard Strasburger's paper calling for the adoption of the phylogenetic method in all aspects of biology.[39] Botanists generally accepted evolution but had little to say about it directly. The theory of descent gave form to the work begun earlier by Hofmeister, but the bulk of botanical research in Germany did not have a direct application to Darwinism. One could work out the details of the production of spores in mosses or the development of vascular bundles in the stems of dicotyledonous plants without recourse to evolution theory. Although eventually he would become a strong critic of natural selection theory, in the 1860s and 1870s Sachs was willing to devote the final section of his textbook to a faithful reproduction of Darwin's argument as it appeared in the early editions of the *Origin of Species*. He spelled out the Darwinian theory in detail, explaining how plants continually produce varieties that differ from each other in minute degrees and stating that these differences offer slight advantages or disadvantages in the continual struggle for existence.[40] In a brief comment at the end of this section, he mentioned Nägeli's belief that the main cause of phylogenetic development is not natural selection but the innate directional variability in plants. Sachs later took a position close to Nägeli's view, but here he simply discussed it in the context of a historical survey of descent theory. In all other respects, Sachs offered in his textbook as clear a statement of Darwin's view on evolution and natural selection as one can find in the botanical literature of the nineteenth century. Considering the broad general influence of Sachs's textbook, not to mention the general popularity of Darwinism in Germany in the 1870s and 1880s, it is not surprising that at least one group of younger botanists training in the 1870s would find inspiration for their work in Darwin's natural selection theory.

BOTANY AS A PROFESSION

The overall effect of Sachs's textbook was to provide a clear and comprehensive theoretical and practical guide to the living plant. Schleiden had established the outline for such a guide in 1842, but Sachs now filled in the details of the outline with the results of twenty-five years of research undertaken by a growing community of botanical professionals. Botany had come of age; it was a science in its own right – if not on a par with physics and chemistry, at least heading in that direction. One sign of its maturity was the proliferation of technical journals. Before 1840 the only botanical journal of wide circulation in Germany was *Flora,* begun by Christian Nees von Esenbeck in 1818. By the early 1880s there were six additional general botanical periodicals and a number of journals of smaller circulation. Hugo von Mohl and D. F. L. von Schlechtendahl began publishing *Botanische Zeitung* in 1843, Nathaniel Pringsheim first published his *Jahrbücher für wissenschaftliche Botanik* in 1858, Adolf Engler's *Botanische Jahrbücher für Systematik, Pflanzenge-*

schichte, und Pflanzengeographie first appeared in 1882, and the recently formed Deutsche Botanische Gesellschaft began to publish its *Berichte* in 1883. In addition, botanists published their works in the reports and proceedings of numerous scientific societies and academies, such as the Akademie der Wissenschaften in Berlin and the Medizinisch-Naturwissenschaftliche Gesellschaft in Jena. To keep up with this outpouring of botanical literature, two review journals came into existence: *Botanischer Jahresbericht* (1873) and *Botanisches Centralblatt* (1880). Both of these listed virtually all of the current botanical literature (inside and outside of Germany) according to subject, offered brief reviews of most of it, and included some original articles.

Most of the botanists who contributed to this literature conducted their research at the universities. There were positions for botanists at some of the agricultural schools and technical colleges, such as the Agricultural Academy at Poppelsdorf or the Agricultural College of Berlin, but these positions usually involved heavy teaching responsibilities and emphasized applied research. Besides, the number of agricultural schools, still on the increase in 1850, was reduced to only a handful by 1880 as agricultural instruction was gradually incorporated within some of the universities, such as those of Halle, Leipzig, and Breslau.[41] However, the number of botany professorships at the universities had not increased appreciably. For his address as rector of the University of Würzburg in 1872, Julius Sachs chose as his topic "On the Present State of Botany in Germany." He complained that botany departments were not adequately staffed to deal with recent changes in the science, and he suggested that each university appoint at least two full professors in botany: one to teach morphology and systematics, and the other to teach anatomy and physiology.[42] His words had little effect. By the 1890s only four of the twenty-one German universities – Berlin, Breslau, Göttingen, and Munich – had two chairs in botany, although enrollments had increased appreciably throughout the university system. Total enrollment in the philosophical faculties of the universities increased from 3,102 in 1850 to 9,295 in 1881 to 15,205 by 1903.[43] There was no shortage of doctoral candidates either, since in the second half of the nineteenth century a university career was one of the preferred and reasonably accessible avenues of social mobility for the middle classes in Germany.[44]

To deal with the increased enrollments, universities could add assistant professors and *Privatdozents* to their staffs, chosen from among the swelling ranks of advanced students, and thus limit the number of full professorships. At Würzburg, for example, the number of full professors increased from thirty-one to forty-four between 1850 and 1903, whereas the number of assistant professors increased from five to eighteen and the number of *Privatdozents* from six to twenty-five. Thus a 42 percent increase in the number of professorships was overshadowed by an almost 300 percent increase in the number of positions below the level of professor during a period in which enrollment in the philosophical faculties at Würzburg increased by over 200 percent.[45] The University of Berlin provided a similar example. There the number of *Privatdozents* grew from 59 in 1850 to 212 in 1903, whereas the number of full professors increased from 57 to 89. The botany faculty reflected this pattern. Berlin added a second professor in botany in 1878, when the rest of the staff consisted of two assistant (*ausserordentlich*) professors and three *Privatdozents*. By 1903, still with two professors, the staff had grown to include four assistant professors and ten *Privatdozents*.[46]

Although the prospects for obtaining a professorship were dim, aspiring young botanists nevertheless came to the institutes, which had been established in nearly all of the universities by the 1870s, and pursued the advanced work that would qualify them for university teaching. After completing their doctorates, they remained at the institutes, sometimes supporting themselves as laboratory assistants. At the proper time – that is, when a postdoctoral student had completed a significant piece of advanced research and when one of the universities was in need of an instructor – he "habilitated" (qualified to teach in a university by presenting and defending a *Habilitationsschrift,* a research paper on some special topic). He then tried to make a living as a *Privatdozent* while awaiting a permanent appointment. Often while he was still associated with a botanical institute at a university, a young botanist found a position at the local *Gymnasium* or at a technical school or agricultural institute. He then waited for a chair to open by the death or retirement of a professor. Those who became assistant professors at least made an adequate living – 3,000 to 5,000 marks by the end of the last century – as compared with the average of 1,500 marks for *Privatdozents*, whose income depended entirely upon fees collected for lectures. Full professors made anywhere from 6,000 to 40,000 marks, with the average lying somewhere over 10,000. Those who achieved the top salary levels were quite well off by the standards of the day; and many young men considered the potential income, and the corresponding prestige, of a professorship well worth the sacrifice of living for years on a substandard income, postponing marriage and family, and generally deferring all sorts of material gratification.[47]

Botany had become a profession in Germany in the second half of the nineteenth century. This was due, in part, to the growth of the universities and the development of research laboratories. However, these institutional trends developed alongside internal changes within the science itself, changes effected primarily by German botanists. In his address at Würzburg in 1872, Sachs could boast: "The advantage which the French and English had over us at the beginning of the century has long since been overcome; our progress has been so rapid that they have not been able to keep pace."[48] The English, he added, still employed the methods and principles developed by Robert Brown forty or fifty years ago; and the French, in botany as in other fields, had arrested the progress of science through overcentralization. Sachs pointed to recent developments in plant science that had shifted the balance toward the German side: improvements in microscopy; its application to cell theory, anatomy, and morphology; and the development of sophisticated physiological studies. He complained that too many German professors still concerned themselves with taxonomy, but even at the time of his address, a large number of the advanced students had already shifted their interests to the newer fields. By the end of the century, Germany was the center for research in cytology, plant morphology, and plant physiology.

Twenty years after Sachs's address, Eduard Strasburger wrote: "During the last half century Germany has attained a prominent place in the field of scientific botany. Good evidence for this lies in the rather flattering state of affairs that foreign botanists take such pleasure in the botanical institutes of the German universities."[49] During the 1870s and 1880s, Sachs's institute at Würzburg and that of de Bary at Strasbourg had been the main choices of foreign botanists. At the time of Strasburger's writing (1893), his own institute at Bonn and Wilhelm Pfeffer's institute at Leipzig had become the favorites of foreign visitors. Botanists in other countries, particularly England and the United States, recognized by the 1870s that there was a kind of botanical instruction that they could obtain only in Germany. When Charles Bessey published his American adaptation of Sachs's textbook in 1880, Harvard botanist Asa Gray commented in a review: "The work concerns itself throughout with what the Germans call 'Scientific Botany' – largely with vegetable anatomy and development, and with particular attention to the lower Cryptogamia."[50] Gray, who devoted his career to the study of the classification and distribution of the flowering plants, may have viewed this work with some ambivalence, but younger botanists, such as Bessey and W. G. Farlow in the United States and F. O. Bower and Francis Darwin in England, embraced these new developments and incorporated them into their own research.[51]

This "scientific botany" was somewhat narrowly conceived, and German botanists realized that their expertise did not extend to all aspects of the study of plants. Strasburger acknowledged that the British maintained their superiority in taxonomic botany throughout the century and that Germans were just beginning to catch up in that department.[52] Sachs admitted in 1872 that "in more recent times German botany has developed less in breadth than in depth," but he added that the general approach of the new plant science had been to determine the essential features of particular phenomena and then build up from there, following the path of physics and chemistry.[53] If German researchers had unlocked many of the secrets of the vital processes of the plant, and of plant structure and development, during the 1850s, 1860s, and 1870s, they had nevertheless restricted their work to their laboratories and asked rather limited questions. For one major group of younger German botanists, however, this situation was to change significantly in the 1880s. Disenchanted with the inductive and piecemeal approach to the study of plants, and excited by the possibilities for biological investigation offered by Darwin's theory of natural selection, these younger men directed their research to the solution of problems having to do with the relationships of plants to their natural environments.

2

Schwendener and Haberlandt: the birth of physiological plant anatomy

Botany in Germany was hardly a monolithic enterprise in the middle third of the nineteenth century. The newer cytological and physiological approaches to plant science existed alongside more traditional taxonomic and morphological studies and a small but active interest in plant geography. Nevertheless, the somewhat narrowly focused analytical approach centered in the laboratory came to dominate German academic botany. From the late 1870s on, however, there was, within the ever-widening circle of professionally trained botanists in Germany, a gradual shift toward problems of a more synthetic nature. A small but articulate segment of the botanical community, represented mainly by younger botanists who received their university training during the 1870s and 1880s, wished to redirect the large quantities of information that had been accumulated in previous decades and to initiate research aimed precisely at gaining an understanding of particular aspects of plant *adaptation*.

Owing to its size, its location, and the fortunate timing of an academic vacancy, one of the most important centers for this movement during the 1880s was the botanical institute directed by Simon Schwendener at the University of Berlin. Situated in Germany's largest university and the capital of an expanding empire, a city replete with ancillary facilities for teaching and research in botany, Schwendener's institute could attract and support a large number of advanced students. Schwendener's role at the institute was that of teacher and coordinator rather than contributor. He brought with him to Berlin a rather thorough training in cytology and plant anatomy and a strong interest in applying that training to matters of broader significance than the description of plant structures. His previous academic appointments had left him little time or opportunity to develop that interest. By the time he came to Berlin, his best research years lay behind him, and the one major work in which he had addressed the problem of adaptation had failed to excite his professional colleagues. His reputation among botanists rested largely on his skill as a microscopist. However, as an inspiring teacher and the director of a large, amply endowed research facility, Schwendener was able to provide in Berlin one of the most appropriate settings in which the newly developed interest in plant adaptation would find its focus. He was aided and encouraged considerably in that enterprise by his fortuitous encounter with Gottlieb Haberlandt, his most able student, just prior to his arrival in Berlin.

SCHWENDENER: FROM BUCHS TO BERLIN

Simon Schwendener (1829–1919)[1] was born into a farming family in the Swiss village of Buchs. As the only son, he expected eventually to take over the family farm; but at his father's urging, he studied science and mathematics at the Academy of Geneva and took a teaching position at the local secondary school. His father hoped that a few years of teaching might serve as useful preparation for a civil service career, but Simon had other interests. When the death of his maternal grandfather left him with a small legacy, he decided to use it to attend the University of Zürich. There he continued his studies in science and mathematics, taking his Ph.D. in 1856. The extreme specialization that was to characterize the universities a decade or so later had not yet arrived; and although Schwendener worked mainly in botany, in order to complete his doctorate he had to pass examinations that covered mineralogy, zoology, physics, and chemistry as well. This broad background in the sciences, especially the physical sciences, would serve him well later in his career.

His dissertation was based upon a comparative investigation of phenological phenomena, a project originally suggested to him by botanist-plant geographer Alphonse de Candolle at the Geneva Academy. Schwendener made a detailed mathematical study of periodic phenomena, such as the time of bud opening and the first appearance of flowers in various plant species, at a number of sites in Switzerland.[2] Although he carried out this work faithfully and diligently, keeping precise records of the various events, this project was too broadly conceived to sustain Schwendener's interest for very long at this stage in his career. His primary goal, now that he had obtained a basic university education, was to become a botanist within the academic tradition that was developing at the German universities. Fortunately, while in Zürich, he met countryman Carl Nägeli, who was then teaching at the recently founded Polytechnic Institute. When Nägeli received a call to the botany professorship at the University of Munich in 1857, he invited Schwendener to come to work as his assistant.

At Nägeli's side in Munich, Schwendener honed and perfected his skills as a microscopist. Nägeli and Schwendener in fact collaborated on a two-volume treatise on microscopy that became a standard laboratory reference manual for two decades. It certainly did not hurt Schwendener's career to have his name linked with that of the premiere plant cytologist of his day.[3] These years in Munich were very productive for him in other ways as well. Here he initiated

Simon Schwendener. Courtesy of the Hunt Institute for Botanical Documentation, Carnegie Mellon University, Pittsburgh, PA. Reproduced by permission of Borntraeger, Stuttgart.

the work that earned him a solid reputation among botanists prior to the Berlin appointment: the identification of the true nature of lichens as pairs of algal–fungal symbionts. Although the complete demonstration of this thesis did not come until years later (through the work of Stahl and Bornet), Schwendener laid the important groundwork by identifying particular algal and fungal cells within the lichen thallus.[4]

Schwendener remained in Munich for ten years, first as Nägeli's assistant and later as a *Privatdozent,* before returning to Switzerland in 1867 to accept the professorship in botany at the University of Basel. There he continued his work on lichens and related studies; but, now free from the influence of Nägeli, he gradually shifted his focus to two larger themes that were to dominate his interest for the rest of his professional life. The first of these, his study of leaf position, was destined to capture the attention of almost no one, yet it remained his pet project. Schwendener was searching for a simple mechanical explanation to account for the spiral pattern in which new leaves appear on a stem, a problem first posed by Goethe. Although he eventually published a book and several articles on the subject, few of his colleagues gave these more than passing notice. The other theme to which he turned while in Basel had far greater influence, especially among younger botanists. This was his application of the science of mechanics to the structural support system of plants. Making use of his background in physics and mathematics, Schwendener examined the relationship between structural tissue, mainly the thickened regions (sclerenchyma) associated with vascular bundles, and the various physical stresses to which monocotyledonous plants are subjected. He published the results of this investigation as *Das mechanische Princip im anatomischen Bau der Monocotylen* in 1874.[5] Although the book itself did not cause an immediate stir in botanical circles, its influence worked indirectly by luring a few younger botanists away from traditional studies.

Schwendener worked in relative isolation in the small Swiss university, in an atmosphere that was pleasant and familiar but hardly stimulating. He was the only botany instructor there, without either assistants or advanced students; his primary responsibility was to teach general botany to undergraduates. Although free to conduct research in what spare time he had, Schwendener missed the stimulation that comes through interaction with colleagues who are pursuing similar interests. After ten years in that environment, he was willing to forsake his native country for an opportunity to enter the mainstream of academic botany. In 1877 he jumped at the offer of the botany professorship at Tübingen to fill the chair vacated by the death of Wilhelm Hofmeister.

The reception of *Das mechanische Princip* by the wider botanical community may have influenced his decision to leave Basel. Schwendener considered his work on the mechanical system in monocots to be the turning point in his research career. It marked his break with the anatomical tradition in which he was trained, a tradition that stressed painstaking description at the expense of interpretation. Although the reviews of *Das mechanische Princip* do not indicate a hostile response, Schwendener was convinced that the indifference, if not resistance, to his ideas was due, in part at least, to his isolation in Basel:

> The reception which *Das mechanische Princip*, appearing in 1874, found with my colleagues was, however, hardly favorable. The two greatest botanical schools in the German Reich, that of de Bary in Strasbourg and that of Sachs in Würzburg, rejected my interpretation; and against such opposition I felt isolated in Basel, without comrades-in-arms and without the hope of training such.[6]

In this excerpt from an autobiographical sketch, Schwendener may have been exaggerating the opposition to his views. For example, Anton de Bary's 1877 comparative anatomy textbook included the following passage regarding *Das mechanische Princip:*

> Schwendener, in his excellent work which has been mentioned so frequently, treated both the arrangement of the sclerenchymatous masses and their physiological relations so minutely and comprehensively that any detailed exposition of the former, within the space here available, must be a mere extract from his work, or it must seem like one.[7]

This hardly sounds like rejection. In fairness to Schwendener, however, de Bary was quite willing to praise thoroughness wherever he found it, but his encyclopedic approach to plant anatomy found little room for the wider implications of Schwendener's work, particularly the concept of a unified mechanical system. Julius Sachs also expressed doubts about the wisdom of calling the network of fibrous tissue that serves as the plant skeleton a "mechanical system." Sachs argued that since sclerenchymatous tissue does not always serve a mechanical function, and since mechanical support is not always provided exclusively by sclerenchyma, it is incorrect to speak of a mechanical system as a physical entity.[8]

Whether at this early stage in his break with the traditional approach to plant anatomy Schwendener's enemies were real or imagined, he nevertheless found an eager and capable ally almost as soon as he arrived in Tübingen. In his words,

> When I received a call to Tübingen, I took it without deliberation, since to me the greater sphere of activity at the Württemburg university offered many advantages. Indeed, soon after my official entrance there I had the pleasure of welcoming a talented young botanist as a student and winning him over to my line of research. This was G[ottlieb] Haberlandt, at that time a young Ph.D. from Vienna, today professor of botany at Graz, and for years my faithful comrade-in-arms in common pursuits.[9]

Haberlandt worked with Schwendener for only one year, 1877–8, returning to Vienna the following year. Meanwhile, Schwendener took advantage of an opportunity to move to Berlin. When Alexander Braun died after serving for twenty-five years as professor of botany at the University of Berlin, two positions were created to replace the one left vacant. August Eichler, a former student of Braun, was appointed to teach systematic botany and to direct the Berlin Botanical Garden and Museum. Simon Schwendener was appointed to teach general botany – which included anatomy, morphology, and physiology – and to direct the newly founded botanical institute. He moved to Berlin in 1878 and remained there until his retirement in 1910.

At Tübingen, despite the presence of Haberlandt, he had not had the time to develop the kind of research program that he had hoped for. For much of that year he was too busy with his writing (he was working on the book on leaf position) to rescue the botanical facilities from the very disorganized state in which the aging and ill Hofmeister had left them. At Berlin the situation was quite different from the beginning. The facilities were much better, and the large number of advanced students both provided a stimulating atmosphere and aided in the teaching of undergraduates. Schwendener's main role in Berlin was to direct the research of advanced students. He himself no longer carried on laboratory research. His study of the mechanical system was to remain his most significant contribution to the school that developed around his botanical institute in Berlin; and although Schwendener's research program found a secure home there, Haberlandt, who did not follow him to Berlin from Tübingen, was to remain his most influential student.

HABERLANDT'S EARLY CAREER

Gottlieb Haberlandt (1854–1945)[10] was born in Hungary and raised in Vienna, where his father, also a botanist, taught at the Agricultural College. Haberlandt attended the University of Vienna and received his doctorate there in 1876 under plant physiologist Julius Wiesner. In his memoirs Haberlandt claimed to have been influenced as much by reading the works of Nägeli, Hofmeister, von Mohl, Franz Unger, and especially Julius Sachs as by the teaching of Wiesner. In the mid-1870s the work of these botanists had not yet filtered down into regular classroom instruction in Vienna. It was largely through Sachs's *Lehrbuch der Botanik,* which was not used in his formal instructional program, that Haberlandt familiarized himself with the newer ideas in botany. Fortunately, his home was well equipped for botanical study: "And under the guidance of Sachs's textbook, I sat down at home at a Merz microscope and observed and sketched much that was passed over in the official practicum."[11]

Haberlandt's early work engaged him in the details of plant anatomy and physiology. At Vienna he produced a paper on cellulose in cork tissue, another on lenticels, and his doctoral thesis on the winter coloration of perennial leaves.[12] However, his interests soon turned to broader themes. In 1877 he published a short treatise on the protective mechanisms of seedlings, a work based upon research that he had carried out at the agricultural school where his father taught. Haberlandt discussed all of the conditions to which young plants must be adapted in order to survive – extremes of temperature, moisture, and light, a variety of soil factors, physical damage, and so on – and he described the various structural and physiological means by which this adaptation is effected.[13] He subtitled this work "A Biological Study," a designation that German botanists in the last quarter of the nineteenth century frequently attached to works of a more general and interpretive nature, especially works dealing with adaptive phenomena.

This was indeed an ambitious undertaking for a twenty-three-year-old botanist barely out of the university. To a great extent, Haberlandt's motivation came from his fascination with the Darwinian point of view; he saw these intricate adaptations as end products of the struggle for existence. In his exuberance, he sent a copy of the book to Darwin, who acknowledged the gift with a brief letter. As Haberlandt tells us: "The letter said that he had learned much from it – a remark that, to be sure, must have been considered safe and meant only as kind encouragement; nevertheless it filled me with happiness and pride."[14] Whatever Darwin's opinion of the book, Anton Kerner's opinion almost won Haberlandt an appointment at Innsbruck in 1878. Kerner, whose ecologically oriented *Pflanzenleben der Donauländer* had made a strong impact on Haberlandt, was leaving Innsbruck to accept an appointment at Vienna.[15] Haberlandt was rather young and inexperienced to be considered for a chair, but Julius Wiesner recommended him highly and Kerner, very impressed by Haberlandt's book, apparently would have been satisfied with the choice of Haberlandt as his successor. However, the Austrian authorities chose the more cautious route and gave the position to the older and more experienced Johann Peyritsch.[16]

Ambitious as his first book may have been, it had little impact on the botanical community apart from Kerner. Haberlandt established his reputation among botanists only in the years following his period of study with Schwendener in Tübingen. In 1877 the Austrian Ministry of Education offered him a year's stipend to study botany anywhere within the German Reich. Wiesner naturally encouraged him to work with Sachs in Würzburg. Such experience, and the prestige carried with it, could not help but benefit the career of a young botanist. A year's work with de Bary in Strasbourg might confer similar advantages. This was, in fact, the precise path chosen by British botanists Sidney Vines and F. O. Bower, who were presented with the same opportunity in the late 1870s. Both chose to work first with Sachs and then with de Bary.[17] Haberlandt, however, decided to spend his year in Tübingen with Schwendener, who, he believed, presented a fresh approach to the study of plant life that would eventually supersede that of de Bary, Sachs, and their contemporaries. In Haberlandt's opinion, Schwendener's *Mechanische Princip* opened up a whole new direction in botanical research by linking the structure and arrangement of plant tissue systems with their physiological functions. Haberlandt discussed his plan with botanist Eduard Strasburger, who was visiting Vienna from Jena at that time. The more cautious Strasburger disapproved, warning Haberlandt that he could not derive much from his contact with Schwendener without having had considerable background in mathematics and mechanics. Haberlandt ignored this warning, and he left for Tübingen, as planned, in the fall of 1877.[18]

His arrival there coincided with the appearance in print of Anton de Bary's new compendium of comparative anatomy, *Vergleichende Anatomie der Vegetationsorgane der Phanerogamen und Farne*. This work consolidated years of careful microscopic research on the vegetative parts of the seed plants and ferns; it was a virtual encyclopedia for the plant anatomist. However, Schwendener took one look at the text and stated simply, "This is a book that is already out of date from the start."[19] Haberlandt realized then that he had made the right decision in coming to Tübingen. To Schwendener, de Bary's book represented just another collection of descriptions, the sort of work that botanists had been engaged in for years. The book simply reported observations without taking into account the relationship between structure and function. Schwendener's work on the mechanical system had at least made an effort to understand that relationship. Such an approach, rather than the monotonous listing of unconnected facts, was to be the model for future botanical research. Haberlandt, needing very little convincing, settled down under Schwendener's tutelage.

The immediate result of Haberlandt's year in Tübingen was the publication of a short treatise on the developmental history of the mechanical system in plants.[20] In his own work on the mechanical system, Schwendener had deliberately ignored the process of development, choosing to confine himself to the physics of the structural tissue as found in mature plants. Considering Strasburger's warning to Haberlandt in Vienna, Schwendener's heavy emphasis on physics and mathematics in a botanical text may, in fact, have contributed to its neglect by the wider botanical community. In any event, the investigation of the developmental history of the mechanical system was the most interesting of the choices for research topics offered Haberlandt by Schwendener, and he applied himself diligently to his task.

From this point on in his career, Haberlandt's work centered on the approach to botanical research that he came to call "physiological plant anatomy"; and he and Schwendener became allies in the struggle to convince their colleagues of the value of this point of view. During most of that time, Schwendener was in Berlin and Haberlandt in Graz. After leaving Tübingen in 1878, Haberlandt spent two years as a *Privatdozent* in Vienna, making use of his work on the developmental history of the mechanical system as his *Habilitationsschrift*. He then taught botany in Graz, in southern Austria, for thirty years. At first he was able to support himself by supplementing his meager income as a *Privatdozent* with a position at the technical college (*technische Hochschule*). Then in 1888 he was appointed to the botany professorship at the University of Graz, a position that he held until 1910, when, largely through Schwendener's influence, he was named as the latter's successor at Berlin. He retired from the University of Berlin in 1924 and remained there until his tragic death during the last year of the Second World War.[21]

THE INITIAL RESPONSE

Most of the criticism for the new point of view fell upon Haberlandt, and perhaps it was this criticism that Schwendener had in mind when he later wrote of the negative reaction to his own work. In contrast to the quiet reception of Schwendener's book,[22] the publication of Haberlandt's *Entwickelungsgeschichte des mechanischen Gewebesystems* evoked harsh criticism from an anonymous reviewer in the prestigious *Botanische Zeitung*, then edited by Anton de Bary. In a sense, this review initiated a controversy between members of Schwendener's circle and the proponents of traditional plant anatomy that was never effectively resolved. On Schwendener's instruction, Haberlandt had set out to determine whether the structural system in plants, a system that functions as an independent unit, is also independent embryologically. With respect to plants, this amounts to determining whether the cells that make up a given tissue originated in the epidermal layer, the nascent vascular strand, or the intermediate region in the growing tip of a shoot. Haberlandt's conclusion was straightforward and unequivocal: "the immediate result of my work is that the mechanical tissue system in plants exhibits no unity embryologically; rather, it is of as many different origins as are possible."[23] His reviewer (perhaps de Bary himself) had no quarrel with this conclusion. Rather, he attacked Haberlandt's motivation for writing the book and the broader implications of his methodology and terminology.[24]

Haberlandt's book was not merely a dispassionate description of plant tissue development; it was an unabashed declaration of a new approach to the study of plant anatomy. Without hesitation, he attributed this new point of view – physiological plant anatomy – to his mentor:

> . . . Schwendener, in his book on the "Mechanical Principle in the Anatomical Structure of the Monocotyledons," provided the first thorough examination of the organization of an anatomical-physiological tissue system. He showed that the arrangement of specifically mechanical cell complexes, particularly the fiber cells, follows entirely different principles from the arrangement of the vascular bundles [with which they are associated], and that where both are similar in structure, this similarity is to be explained not as morphological necessity but as the result of certain physiological advantages.[25]

Thus structural tissue, in this case the fibrous cells associated with the vascular system, follows its own organizational principles, regardless of embryological origin. There was nothing complex or mysterious about the organizational principles; Schwendener had in mind merely the shape and orientation of the cells. Yet Haberlandt went a step further and made explicit what Schwendener had only implied in his cautious exposition: Such cell complexes, Haberlandt insisted, represent integral anatomical-physiological *systems,* the investigation of which promises to offer valuable insights into plant structure that cannot be gained from a strictly anatomical approach.

What had excited Haberlandt most about his examination of the mechanical system was the striking anatomical uniformity in various examples of fibrous tissue, despite its multiplicity of origins. That anatomical uniformity can be explained, he proclaimed, "only on the basis of a uniformity of functions, which proves, in this case, to be a sharper classification principle than any that developmental history or morphology can offer us."[26] For Haberlandt the structural support tissue in plants offered clear evidence demonstrating the necessity of integrating functional considerations with anatomical descriptions; it made no sense to separate the two. The reviewer in *Botanische Zeitung* objected strongly both to this view and to Haberlandt's attributing it to Schwendener:

> One might suppose that the author had reproduced the essence of Schwendener's book. The referent deems it his duty, therefore, to state that in the latter the expression "anatomical-physiological tissue system" occurs about as infrequently as, perhaps, a "sharper classification principle" is sought there in the "uniformity of function." Also, Schwendener does not speak of a "mechanical tissue system"; rather, he interprets "similar cell forms which make up an anatomical system designated by particular characteristics" as "specifically mechanical," and consequently he explains the arrangement of this anatomical system. On the basis of the morphological uniformity found in the fiber system, he then arrives at the purely physiological notion of a "mechanical system."[27]

Here in simplified form was the fundamental objection raised against Haberlandt's work. The reviewer could accept a purely physiological discussion of the role of fibrous tissue in the structural support network of the plant. He could not accept Haberlandt's treatment of this tissue as part of a system that derives its identity from *both* anatomical and physiological considerations. Moreover, Haberlandt had gone a step further, claiming that similarity of function provides a better basis for classification of plant tissues than any that can be derived from the study of morphology or development. The reviewer found this concept entirely unacceptable. A classification system based upon function makes little sense, he argued, because it is so often the case that individual cells take on multiple functions. By insisting on describing physiological-anatomical systems as real entities, Haberlandt has created a cumbersome pseudo-terminology. The review ended with words of caution and a not so kindly reference to Haberlandt's linguistic skills:

> . . . and finally we should express the wish that in the future the author be less rash in his eagerness to adopt "new classification principles," "new concepts," and "new words," especially, as occurred this time, with regard to thoroughly valuable existing terminology. Lastly, the referent asks the author, in the interest of "our beloved German," in the future no longer to write *am* where it must mean *auf dem.*[28]

Haberlandt may have taken this grammatical advice to heart, but he did not waver in his commitment to a new botanical research program with a new terminology.

The preceding review appeared in the spring of 1879. Slightly over a year later, Simon Schwendener was formally inducted into the Berlin Academy of Sciences. He took this occasion to explain the trajectory of his scientific career since the time of his association with Nägeli:

> . . . I was no longer content with merely descriptive anatomy and developmental history. I needed to pursue microscopical studies on a deeper level; and I tried, therefore, to find the governing principle among details relating to the structure and arrangement of particular tissues. I believe that, in this

> way, I described – and, as result, correctly identified according to structure and function – one of the most distinct anatomical systems, namely that upon which the strength of vegetative organs depends, as a construction developed according to the principles of mechanics and adapted to external living conditions. This was, to be sure, but a small step toward a distant goal. What I have in mind is a physiological-anatomical investigation, carried out in analogous fashion, of all tissue systems. . . . In a certain sense, then, this is to be a physiology of tissues, one which doubtless must be accomplished through serious effort, but also one which would serve to restore and revive our lifeless systems of anatomy through the elucidation, in a manner more in conformance with nature, of the relationships between structure and function.[29]

What Schwendener described at the end of this passage is precisely the project carried out by his students, above all by Haberlandt. Whether or not the critical response to Haberlandt's work motivated Schwendener to describe his own goals in these terms, he clearly saw no need to make a distinction between his more cautious approach to his subject, as characterized by *Das mechanishe Princip*, and Haberlandt's bold pronouncement of a new physiological-anatomical method in *Entwickelungsgeschichte*. The passage quoted previously does not commit Schwendener completely to Haberlandt's point of view, but it does indicate his commitment to a research program centered on physiological anatomy. In any event, Schwendener left to his young protégé the full exploitation of this new approach to plant anatomy.

PHYSIOLOGISCHE PFLANZENANATOMIE

Undaunted by the attack in *Botanische Zeitung*, Haberlandt spent the next five years (the first two in Vienna and the next three in Graz) refining his views and examining other plant tissue systems from the perspective of physiological anatomy. In 1881 he published a lengthy article on the assimilation (photosynthetic) system in plants. In the following year, his essay on the physiology of plant tissues appeared in August Schenk's *Handbuch der Botanik*. By 1884 he had expanded this essay into the book that would become his most influential work, *Physiologische Pflanzenanatomie*.[30] Writing in each of these instances with an excitement that sometimes gave way to polemics, Haberlandt, not yet thirty, left the reader with the impression that he was helping to extricate botany from the once useful, but now cumbersome and unnecessary, restrictions of an outmoded anatomical tradition.

The central message of Haberlandt's program of physiological plant anatomy was that accurate and complete explanations for structural phenomena in plants require a complete understanding of the roles of the various structures in the living plant. This was certainly not a novel idea in itself. Few, if any, of Haberlandt's contemporaries would have denied its value in theory, but most would have doubted its practical value. Haberlandt, however, wished to develop this concept into a methodological principle. In his 1881 essay on the assimilation system, he summarized his approach as follows:

> Every morphological feature, may it exist in the form of a particular cell or in the structure of its walls, in the organization of tissue, in the position of an organ, etc., is the result of particular (partly internal, partly external) factors which, in the last analysis, can be identified as chemical and physical forces. However, for the investigation of living beings, or organisms, we cannot be satisfied with a purely mechanistic explanation of morphological facts. At this point [in our investigation] we do not always know the final purpose of these occurrences, i.e., we do not know the significance of every fact under consideration or the role it plays in the life of the organism. The investigation of its physiological role, therefore, is the next matter that we must resolve; but it is not the last. It yet remains for us to demonstrate that the morphological fact is in agreement with the physiological function, i.e., that its characteristic features serve to support and promote the execution of its physiological function.[31]

Thus, wishing to go beyond mere description and beyond physicochemical reductionism, Haberlandt attempted to integrate completely the methods of the physiologist and the anatomist around the central concept of function. He did not want simply to discuss the physiological aspects of known anatomical systems; he wanted to identify the systems themselves, from the beginning, as structural-functional wholes.

In *Physiologische Pflanzenanatomie* Haberlandt identified seven such systems: dermal (the outer skin), mechanical (structural support), absorption (roots, etc.), assimilation (photosynthesis), vascular, storage, and ventilation. He discussed each of these systems in turn, beginning each discussion with a description of the component cells and tissues of the respective system and then outlining the main organizing principles operating within that system. The latter were always variations on one or more of the following:

1. the principle of division of labor
2. the mechanical principle (every detail of construction must satisfy the laws of mechanics)
3. the principle of economy of material (the greatest result must be obtained with the smallest expenditure of material)
4. the principle of maximum exposure of surface
5. the principle of efficiency.[32]

These organizing principles clearly reflect an effort to link botany with the physical sciences, particularly energy physics. Haberlandt was obviously influenced by the formulation of the laws of thermodynamics in the mid-nineteenth century, just as he was influenced by a strict interpretation of Darwinism, which dictated that all structures must have an adaptive function and

Gottlieb Haberlandt. Courtesy of the Hunt Institute for Botanical Documentation, Carnegie Mellon University, Pittsburgh, PA. From *Phyton* (Horn, Austria) 6 (1955): 2/3.

that, within the limitations of a particular biological form, only those structural variations will survive that ensure the execution of their respective functions in the most efficient manner. Darwinism, so interpreted, offered the necessary connection between physical theory and organic structure and provided, therefore, a justification for the development of a *physiological* approach to plant anatomy. Although he had to modify much of the contents over the years, Haberlandt maintained the same approach to his subject and the same general organization of the text in each of the five subsequent editions of *Physiologische Pflanzenanatomie,* the last of which appeared in 1924, the year he retired from Berlin.[33]

The centerpiece of the book, from the first edition on, was the photosynthetic system. Photosynthesis is, after all, the activity most peculiar to plants. It was the subject of Haberlandt's first work devoted entirely to physiological plant anatomy (the 1881 article), and it remained the subject of much of his later research. The word "photosynthesis" had not yet come into general use in the 1880s. Haberlandt, following the lead of Sachs, called the process "assimilation" (for carbon assimilation), restricting the meaning of that term, as did Sachs, to the production of organic matter from carbon dioxide and water, with the consequent release of oxygen. To avoid confusion with the current usage of "assimilation," however, I will use "photosynthesis" and "photosynthetic system" in the following discussion.

Haberlandt's treatment of the photosynthetic system focused on the layered, chlorophyll-rich tissue lying just beneath the surface of leaves. His predecessors had employed a variety of schemes to classify this tissue. In an effort to inject some order into the diversity of cell types that confronts the microscopist, Julius Sachs had included the photosynthesizing cells within the general category "fundamental tissue" (*Grundgewebe*), a term applicable to all tissue that is neither "dermal" nor "vascular."[34] This tripartite distinction – skin tissue, conduction tissue, fundamental tissue – is readily evident to anyone who examines a longitudinal section of a growing tip of a stem or root under the microscope; and Sachs, who was very interested in the process of growth and differentiation but relatively unconcerned with the elaboration of tissue classification schemes, saw no need to subdivide his categories further. Anton de Bary, on the other hand, saw little value in Sachs's tripartite classification. He insisted instead on a more elaborate scheme based upon purely structural, and not developmental, considerations. He identified six types of tissue in the mature plant, with numerous subdivisions within each type. According to this scheme, photosynthetic tissue fell into the category "cellular tissue," that is, tissue characterized by distinct types of living cells (as opposed to that which is characterized by components or remnants of cells, such as sclerenchyma, tracheae, or sieve tubes). Within this general division, photosynthetic tissue fell into the subcategory "parenchyma," a term that applies to a variety of tissues composed of thin-walled, polyhedral cells. Of special interest for the photosynthetic system is the *pallisade parenchyma,* the chlorophyll-rich tissue located just beneath the upper surface of the leaf, where most of the photosynthetic activity takes place, and the *spongy parenchyma,* loosely packed, irregularly shaped cells lying beneath the more regular and densely packed cells of the pallisade layer. De Bary's comparative anatomy textbook included elaborate discussions of the subtle differences in the arrangement of pallisade and spongy parenchyma from one type of plant to another; but the reader will search in vain for any explanation of these differences – they are merely cataloged.[35]

Against this background, Haberlandt's discussion of the photosynthetic system must have seemed at once refreshing, ambitious, and somewhat heretical. He agreed with Sachs that three general categories for plant tissue are sufficient, but he had no desire simply to lay out this general structural scheme and then move on, as did Sachs, to a discussion of the physiological processes that occur within particular regions of the plant. Still less did he wish, following de Bary, to describe the photosynthetic tissue as a subdivision of a subdivision of a categorical tissue scheme based solely on structural features. Instead he identified a general photosynthetic tissue *system,* composed of pallisade and spongy parenchyma, the adjacent cells, and the network of conducting tissue whereby the photosynthetic products are transferred out of the region of their manufacture. He then argued that all variations on this general scheme, as they manifest themselves in particular groups of plants, represent adaptive responses to specific physiological needs, in conformance with (1) the principle of maximum exposure of surface area and (2) the principle of expeditious translocation, that is, the removal of synthetic products by the shortest and most efficient route. The first principle, according to Haberlandt, accounts for the branched and loosely aggregated cells found in the spongy layer of most higher plants. This arrangement provides ample surface area, as well as an abundance of air spaces, to facilitate gas exchange to the cells in the adjacent pallisade tissue. The second principle – based upon the general rule that chemical reactions proceed most efficiently when their products are quickly removed – accounts for the gradual development within the plant kingdom of specialized transport mechanisms. Haberlandt saw a clear evolutionary trend, from the simplest arrangement, found in mosses and aquatic plants, whereby each photosynthetic cell serves also to transport its own products, to the most complex arrangement, found in grasses and many dicotyledonous plants, whereby synthetic products pass from the specialized cells where they are manufactured through special intermediary tissues before removal to other regions of the plant. After laying out these general principles, Haberlandt identified several possible arrangements of photosynthetic tissue, in order of increasing sophistication, and described numerous examples representing each of these arrangements. He offered elaborate descriptions of the types of cells found within the photosynthetic system, he discussed the nuances of chloroplasts at great length, and he gave considerable attention to physiological detail.[36]

The other systems received comparable treatment. This attention to detail, along with Haberlandt's command of the literature and his obvious firsthand familiarity with his subject, gave the book considerable respectability. However, this was clearly not another text written in the inductive tradition of German laboratory botany. What set Haberlandt's work apart, and drew some criticism, was his attempt to abstract from the deluge of detailed anatomical and physiological information applicable to each system the few organizing principles that give the system its integrity and account for its variant forms throughout the plant kingdom. This kind of bold synthesis, making use of the generalizing power of the physical sciences as well as evolution theory, required a leap of faith. Haberlandt was well aware that all of the evidence was not in:

> So I resolved to take a daring step that only the "courage of youth" could justify: [I decided] not to wait until such an abundance of material had been assembled in a string of unconnected works that a summary in a large work would fall as a ripe fruit from a tree. Decades would have to elapse for that. After careful deliberation, I made up my mind to set up the general framework in which the particular systems are housed.[37]

But eschewing the inductive method required more than impatience and the courage of youth. It required commitment to a view that plant structures are comprehensible only within a functional context. Once one is committed to this view, it makes more sense to talk about physiological-anatomical systems than to elaborate anatomical classification schemes and talk later about the physiological roles of some of the anatomical structures.

Haberlandt was not alone in his desire to explore more closely the relationship between structure and function, but no one had yet attempted to bring together in one volume the scattered fragments of work relating to physiological plant anatomy. In brief historical remarks in the fourth edition of *Physiologische Pflanzenanatomie* he once again credited Schwendener with providing the stimulus for this synthesis:

> The earlier work in this field, however, produced little more than a mass of disconnected observations, which admitted only of the vaguest generalization; no methodological and exhaustive account of the connection between the structure and functions of any tissue-system had been written previous to the year 1874, when Schwendener published his classical treatise on "the Mechanical Principle Underlying the Anatomical Structure of the Monocotyledonous Plants," in which the definition and diagnosis of a tissue-system was for the first time carried out and consistently in accordance with principles now generally accepted by physiological anatomists.[38]

Schwendener's accomplishments notwithstanding, Haberlandt went much further by treating *all* plant systems in this way and by quite deliberately presenting his work within the context of a new methodology, one that was grounded in a Darwinian approach to plant adaptation.

3

Overtures to Darwinism

Looking back from the perspective of the late twentieth century, a century in which functionalism in its various forms has played so large a role in the biological and social sciences, it is somewhat difficult to understand fully why Haberlandt's functional approach to plant anatomy caused such a stir in the 1880s. It was hardly the central controversy in the life sciences in the late nineteenth century, and Haberlandt, like Schwendener, may well have exaggerated the opposition to his views, but the opposition did exist, and it persisted to the end of the century. Opponents of Haberlandt's approach did not deny the functional significance of particular plant structures, they simply disagreed with Haberlandt's view that structure is best studied in terms of function. They insisted that structure must be studied in and of itself. This mainstream view of the study of structure can be viewed as a kind of hard-headed empiricism, but it also had its roots in the German morphological tradition, the pure study of organic form articulated so well by Goethe at the end of the eighteenth century.[1] Haberlandt, who was hardly unaware of the debt that botanical science owed to Goethe, was convinced that Darwin's natural selection theory justified his break with that tradition.

He was also very much aware of his break with more recent botanical traditions. He continued to view his work as representing a radical departure from the prevailing approaches of de Bary and Sachs. He later wrote of *Physiologische Pflanzenanatomie*: "While Schwendener's school applauded it joyfully, it met with unanimous rejection on the part of the disciples of pure descriptive anatomy under the leadership of Anton de Bary."[2] These botanists, he added, were ruled by the maxim that anatomy and physiology must not be treated together. He was no less candid regarding the divergence of his views from those of Sachs. When botanists in the early twentieth century identified Sachs as one of the founders of physiological plant anatomy, Haberlandt felt compelled to publish a short note to set the record straight. He stated that Sachs belonged quite clearly on the side of the purely descriptive plant anatomists. While acknowledging the brilliance of Sachs's *Lehrbuch*, Haberlandt stated bluntly that insofar as physiological plant anatomy was concerned, "I did not draw the slightest immediate stimulation from this work."[3] He did draw immediate stimulation from Darwin's work, however, and he did not hesitate to frame his functional anatomy within a Darwinian context.

NEW CRITICISMS

Haberlandt's ambitious synthetic work brought forth the expected response in the pages of *Botanische Zeitung*. Otto Warburg, a student of de Bary, gave *Physiologische Pflanzenanatomie* a mixed review, praising its thoroughness but criticizing its lofty goal. Far from achieving a synthesis of anatomy and physiology, Warburg claimed, Haberlandt has created simply another branch of physiology, one that considers structural features from a purely physiological point of view. Such an enterprise is perfectly acceptable, Warburg argued, as long as the author leaves no doubt in the mind of the reader as to what he is doing. Haberlandt, however, creates a terminology that leaves the reader with the false impression that anatomical-physiological systems actually exist in nature. Further, since Haberlandt chooses to deal with questions involving purpose while claiming to deal with tissue systems, his method could better be termed "teleological anatomy" than "physiological anatomy."[4]

Haberlandt was not disturbed by such remarks. He knew that any discussion of the interrelationship between structure and function could not avoid teleological language. The pages of *Physiologische Pflanzenanatomie* are filled with discussions of purpose. Although Haberlandt almost matches de Bary in terms of quantity of anatomical detail, his discussions are quite deliberately centered on explanations of that detail. Attempting to *explain* structural details rather than describe them places one on shakier ground; and although Haberlandt tried to confine his explanations within the limits of available anatomical and physiological knowledge, many of his professional colleagues simply rejected outright any attempt to make conjectures as to the purpose of this or that structure as a basis for describing and classifying structure.

According to Haberlandt's account, some of the informal criticism leveled at him in classrooms and at scientific meetings was considerably less restrained than that which appeared in print.[5] De Bary allegedly held up a copy of *Physiologische Pflanzenanatomie* before the students in his botanical practicum in Strasbourg and declared, "Here you have the latest botanical fiction." Bacteriologist Ferdinand Cohn approached Haberlandt at a meeting to tell him that he had locked up his copy of the book as soon as he received it in order to keep it from his students. Haberlandt replied that it gave him pleasure to be considered so dangerous a heretic. At the same meeting, paleobotanist H. Graf zu Solms-Laubach ridiculed Haberlandt, in an otherwise serious scientific presentation, by describing the host plant of a parasitic fungus as, according to Haberlandt's point of view, a "fungal-spore-dispersal-tissue-system."[6] Solms-Laubach, along with many of his contemporaries, assumed that any tissue classification scheme based upon function could not fail to result in such ludicrous categories. Haberlandt had attempted to answer such objections in his book, both in his introductory remarks and in the sections dealing with particular systems. Every cell containing chlorophyll, he argued, is not necessarily to be included in the photosynthetic system; every cell that stores nutrients is not necessarily part of the storage system. Inclusion within these systems depends upon the primary function of the cell, its location, and its relationship to other cells within the given system. It is not always easy to determine whether or not a particular cell, or group of cells, should be included within a particular system, but Haberlandt suggested following simple pragmatic guidelines. For example, in the case of the photosynthetic system, he suggested that the percentage of chlorophyll content should serve as an accurate criterion for inclusion.[7] The important matter for Haberlandt, in any event, was to treat structural features within the context of their functions in the life of the plant rather than to set up rigid boundaries for the respective systems.

Haberlandt found considerable justification for this teleological mode of explanation in Darwin's evolution theory. In the first edition of *Physiologische Pflanzenanatomie,* in the general introductory remarks regarding the physiology of plant tissues, he pointed out the difference between mechanistic and teleological explanations. The former concerns the efficient cause of a structural feature or set of features. Until recently, he argued, biologists have rested their interpretations on a solid footing only when they have restricted themselves to a discussion of efficient causes, since there has been no satisfactory means of incorporating final causes within the framework of their science. All of this has changed, however, as a result of Darwin's work:

> In the "struggle for existence" only those morphological variations that guarantee the safest, most complete, and most efficient operation of all physiological functions become fixed through inheritance; or, to put it more precisely, that combination of chemical and physical forces that brings about, in a causal-mechanistic fashion, the advantageous morphological characteristics in every individual of the species in question will persist through inheritance. Thus efficient causes are linked to final causes; the former assures, in the development of every individual, the appearance of particular morphological features; the latter assures the same in the historical development of the entire species, genus, or family.[8]

Although Haberlandt's contemporaries were certainly familiar with this interpretation of Darwin's theory, they generally omitted such discussions from their botanical work. For Haberlandt, however, Darwin's theory offered the justification for treating structure and function together. Haberlandt considered the static classification of tissues, based upon purely structural considerations, to be wholly inadequate. The proper subject for physiological anatomy is not a particular structural system but a *causal nexus* produced by morphological features in conjunction with physiological processes. This causal nexus neither remains static nor arises anew with each individual; instead it "is formed . . . during the development of the entire species and, indeed, very gradually, through adaptation by means of natural selection."[9] Thus the subjects of physiological plant anatomy, the integrated systems, are not distinct physical entities at all but abstract conceptions derived from viewing the close relationships between structure and function within a strict Darwinian evolutionary perspective.

For a botanist to base a textbook concerning anatomy and physiology on Darwinian principles was unique in 1884. The concept of natural selection is difficult to incorporate into a discussion of cells, tissues, and physiological processes, especially for scientists trained to report observations and avoid interpretations. Schwendener certainly made no mention of natural selection in his work, and Haberlandt hints in the following passage that he and his teacher did not share the same respect for Darwin:

> In the first edition of my book, I took my stand entirely on the basis of Darwin's natural selection theory, and in it I caught sight of the key to the mechanistic explanation of useful structural and organizational adaptations. Schwendener . . . brought to my attention at that time his belief that a Darwinian train of thought is not instrumental in the discovery and description of mechanical systems [in plants].[10]

Schwendener's attitude did little to dampen Haberlandt's enthusiasm for Darwinism. The ubiquity and efficacy of natural selection were implicit assumptions throughout *Physiologische Pflanzenanatomie*.

DARWINISM IN GERMAN BIOLOGY

The skepticism that Schwendener expressed toward Darwinism was not unusual for botanists and zoologists of his generation. Due to the nature of German biology since the 1840s – its strong laboratory orientation, its mechanistic emphasis, its divorce from natural history – even biologists who were quite willing to accept evolutionary change as a fact of nature nevertheless found Darwin's causal explanation to be inadequate.[11] The natural selection theory left too much to chance; it was too messy and open-ended to satisfy a group of professionals used to probing the microscopic details of organic structure and the nuances of organic processes. Skeptics like Schwendener shared this assessment of natural selection with some of the stronger critics from an earlier generation. Perhaps the embryologist Karl Ernst von Baer summed up the central issue best in his somewhat belated critique of the Darwinian theory:

> I too am convinced that everything that exists and continues to exist in nature arose and will continue to arise through natural forces and material substances. But these natural forces must be coordinated or directed. Forces which are not directed – so-called blind forces – can never, as far as I can see, produce order.[12]

Von Baer's conviction that the wonderful order observable in organic nature cannot be the result of the vicissitudes of blind chance may have been grounded in religious belief, but it was also based upon a particular interpretation of the order and regularity of natural phenomena. According to this view, orderly phenomena in nature must always be attributed to regular, repeatable causes. The kind of order that is represented by the total array of living and extinct organic forms delineated by our various classification schemes cannot be explained as the product of an undirected force acting upon variation brought about by chance. This view was expressed by critics of Darwin in England, France, and the United States as well, but it had a particularly strong meaning within the framework of German life science in the mid-nineteenth century. A generation of biologists like Hofmeister, Nägeli, and Schwendener, individuals who had helped to establish a laboratory-oriented, inductive approach to the life sciences based upon cell theory and embryology, found it extremely difficult to attribute the gradual unfolding of organic diversity to the willy-nilly actions of natural selection.

Not surprisingly, the German Darwinists, that is, those who recognized a central role for natural selection, were often field-oriented scientists, such as Moritz Wagner, Fritz Müller, Carl Semper, and Anton Kerner. It was easier for field scientists used to observing both the variability of organic productions in the wild and the diversity and severity of external conditions to grasp the meaning and significance of natural selection. Darwinists also tended to be younger men who, in the early 1860s, were not yet established in their respective fields. Thus Fritz Müller (1821–97), who wrote one of the earliest works supporting Darwin's natural selection theory, was living in self-imposed exile in Brazil following the failure of the 1848 revolutions. His brother Hermann (1829–84), a botanist who examined the interrelationships between flowering plants and their insect pollinators, remained in Germany to teach biology at a *Realschule* in Westphalia, eventually getting into trouble with the Prussian authorities for presenting Darwinian ideas in the classroom.[13] The zoologists Ernst Haeckel (1834–1919), Karl Semper (1832–93), and August Weismann (1834–1914) were barely beginning their academic careers in the early 1860s. The same could be said for Julius Sachs (1832–97), who was teaching at an agricultural school in Poppelsdorf, and Anton Kerner (1831–

98), who had just begun his first university appointment at Innsbruck. Each of these individuals, all born within three years of each other, embraced the Darwinian theory more strongly than biologists in the generation immediately preceding theirs (although, as mentioned earlier, Sachs later became a strong critic of natural selection).[14]

The popularity of Darwin's version of evolution theory among younger scientists had to do with more than the merits of the theory itself. In Germany, Darwinism came to symbolize the rejection of traditional religious beliefs and values, the rejection of traditional forms of political authority, and progressivism in general, whatever problems may have existed in the actual application of Darwin's work to these cultural and social issues.[15] Some of the strongest proponents of Darwinism, such as Haeckel, the brothers Müller, and Weismann, were political liberals in favor of social reform and the unification of the German-speaking nations. Haeckel incorporated his political views into his *Generelle Morphologie* (1866), the first pro-Darwinian treatise to receive wide attention in Germany, and his considerably more popular *Natürliche Schopfungsgeschichte* (1868). In the former work, he made a case for a unified political state analogous to the organism as a unified and integrated collection of independent cells, the product of Darwinian evolution.[16] Haeckel was a complex figure by no means representative of all German Darwinists. Yet whatever opinion other young Darwinists had of Haeckel's political analogies, many shared his view that the Darwinian theory could account for the diversity of organic productions without an appeal to directing forces of any sort; and they shared his belief that the key to understanding this process of diversification was the comparative study of embryological development grounded, in principle at least, in cell theory.

Older and established zoologists and botanists tended to view Darwin's evolution theory as irreconcilable with the idealistic approach to morphology that, despite newer developments in methodology and technique, still held considerable influence. Objections came from two corners. Those who opposed the idea of organic transformation in general argued that the type concept precluded any modifications in structure over time. The notion that all organisms can be grouped into a limited number of stable types implied for many that each type is characterized by its own unique arrangement and integration of parts. For those who subscribed to this view, modifications in form, no matter how slight, could not fail to be detrimental to the individual organism, and transformations from one type to another were totally inconceivable. This had been the view of Cuvier as he reacted against French transformist theories earlier in the century. Unfortunately for Darwin, this was a view shared, in part at least, by Heidelberg paleontologist Heinrich Bronn, who undertook the first German translation of the *Origin of Species*. Because of his strong objections to Darwin's theory, Bronn felt obliged to append his own critical closing remarks to the German text.[17] Due to Darwin's objections to the Bronn translation and to the intervening publication of pro-Darwin works in Germany, especially Haeckel's *Generelle Morphologie,* later translations were done by the more sympathetic Viktor Carus.[18]

Others working within this idealistic tradition accepted evolution in theory while rejecting Darwin's proposed mechanism. After all, theories of organic transformation had appeared in German biology in the first half of the nineteenth century, and Darwin offered the most plausible evidence thus far in support of the general notion of descent with modification. Natural selection, however, was another matter. Although younger proponents of Darwinism might welcome natural selection as the last link in a completely mechanistic understanding of the universe, biologists with one foot in the idealistic tradition could not accept such a seemingly unlawlike explanation for species change. They were used to viewing scientific laws in terms of either simple linear sequences of cause and effect or general mathematical formulas.

Carl Nägeli's views toward evolution exemplify this second position. In 1865 he argued that a phenomenon as complex as the production of organic diversity could not have been the result of natural selection acting upon chance variation. At best, Nägeli insisted, natural selection serves as a pruning device by eliminating the least fit varieties; but the creative role in evolution is played by internal forces that are responsible for promoting directional variation independent of external factors. Nägeli argued that since many organic structures do not serve an adaptive function, we can only assume that there are internal mechanical forces that direct organic change along certain paths.[19] The question of the usefulness of organic structures remained a matter of serious debate well into the twentieth century, since if internal mechanisms are capable of continually generating nonadaptive (or neutral) characters, then the attention of evolutionists should be focused more on the internal mechanisms themselves and less on the relationship of organisms to their external environments. Nägeli returned to this theme in a later work (see Chapter 9), and zoologist Theodor Eimer popularized the general view of nonadaptive linear evolution as "orthogenesis."[20]

However, in the first two decades following the publication of the *Origin of Species,* the debate over the efficacy of natural selection was overshadowed by interest in "phylogeny," a word coined by Ernst Haeckel in his *Generelle Morphologie* of 1866. Many botanists and zoologists expended their energies on the construction of plausible genealogies and ignored, for the time being, the question of adaptation. Biologists who had once been steeped in the idealistic tradition, such as Carl Gegenbaur, now used the methods of comparative anatomy to establish plausible lines of descent.[21] The search for unity in the organic world, a theme that in the first half of the nineteenth century amounted to a preoccupation with an idealistic unity based upon affinities in design and/or supplemented by a belief in a progressive series of organisms from simplest to most complex, now focused on a material unity based upon common descent. The construction of phylogenetic schemes became an obsession with many zoologists, but it was also one of the central interests of German plant taxonomists and plant geographers.

Although interest in the general problem of phylogeny continued to the end of the century, attention began to return to the mechanism of evolution in the 1880s as the work of Oscar Hertwig, Eduard Strasburger, and especially August Weismann pointed to the central role of the cell nucleus in hereditary transmission.[22] Weismann's theory of the continuity of the germ plasm, first formally proposed in the early 1880s, engendered a debate between strict Darwinians and neo-Lamarckians over the importance of natural selection in the production of organic diversity. If, in fact, there is no connection whatever between the germ plasm and the somatic cells, the Darwinists argued, then natural selection must be the central causal agent in organic transformation. The reaction to Weismann's cytological theory was strong, however, and the 1890s saw an increase of interest among German biologists in the direct role of the external environment in inducing adaptive variations. Whatever position one took on this matter, Darwinian or neo-Lamarckian, the focus was on the role of the environment, whether direct (through the inheritance of acquired characteristics) or indirect (through natural selection acting upon chance variation), in effecting evolutionary change.

THE NEW DARWINIAN BOTANISTS

The botanists under consideration in the present study received their university training and began their active research in the late 1870s and the 1880s, when the debate over the alternative environmental evolution theories began to take shape. Schwendener, who was a young *Privatdozent* working under Nägeli when Darwin's *Origin of Species* appeared in print, took a position close to Nägeli's, arguing that internal forces, rather than external conditions, are the creative agents in evolutionary change.[23] Haberlandt, however, was introduced to Darwin's views in his youth by reading the popular works of Haeckel.[24] He and his contemporaries were among the first generation of German biologists to come of age as scientists entirely within an evolutionary context. Although there was still considerable debate surrounding the teaching of Darwinism in the public schools, there was little serious opposition to evolution theory among German university scientists in the 1870s. Moreover, although most German evolutionists were not strict Darwinists, interest in the natural selection hypothesis was at its peak in the 1870s and 1880s.

Whereas Schwendener, the cautious empiricist, maintained a critical attitude toward natural selection, Haberlandt, an ambitious youth inclined to side with the much criticized Haeckel and searching for a theoretical basis for a synthetic approach to plant anatomy and physiology, embraced Darwinism wholeheartedly. Darwinism must be viewed as the intellectual context in which Haberlandt presented his particular discussion of adaptation. Whereas Schwendener treated adaptive phenomena in a very mechanical fashion, choosing to confine himself to the details of the particular process at hand, Haberlandt believed that Darwin's natural selection theory gave him free reign to discuss adaptation within a broader context. In practice, he did little more than state that a particular structure or arrangement had a selective advantage over other types of structure or arrangement; he made no attempt to trace phylogenetic histories of particular structures and correlate these with knowledge of past environmental changes. Such studies still present immense difficulties for evolutionary biologists. However, the difference in point of view between Schwendener and Haberlandt represented an important, although subtle, shift in focus. Against his own training and a thirty-year trend in German botany, Schwendener had dared to reintroduce a discussion of adaptation into the study of plant anatomy, and his example provided the impetus for Haberlandt's work. One can discuss adaptation without evolution, or course, but within an evolutionary context adaptation takes on a dynamic aspect. Haberlandt not only introduced the evolutionary context, but by emphasizing evolution by means of natural selection he directed attention even more to the external environment of the plant.

The shift in focus from the laboratory to the natural setting of the plant characterized much of German botany in the past quarter of the nineteenth century. The mechanistic-reductionistic program of the 1850s, 1860s, and 1870s that had begun with Schleiden was perhaps losing its attraction for younger botanists, who felt the need to synthesize the various unconnected strands of anatomical and physiological research that had developed during those years. Darwin's evolution theory offered the theoretical basis for such a synthesis. On the one hand, it demonstrated the phylogenetic unity of all plant life; on the other hand, it presented biologists with the opportunity to focus the techniques developed in the laboratory on the problem of adaptation. Also, as the following chapters will make clear, Germany's acquisition of overseas colonies provided an additional stimulus for the study of adaptation by offering botanists the opportunity to experience novel environments. Within this context, the work of Schwendener and Haberlandt provided botanists of Haberlandt's generation with a framework and a formal justification for further research. Haberlandt himself had little direct influence on the development of the new school of research, since at the University of Graz he was hardly in a position to amass a large following. *Physiologische Pflanzenanatomie* remained his most significant contribution to the new approach to the study of plants – an approach that, through the efforts of Schwendener, became the dominant theme of a group of young botanists at the University of Berlin during the 1880s.

4

Schwendener's circle: botanical "comrades-in-arms"

Whereas the study of plant adaptation found a number of homes within the German botanical community in the late nineteenth century, the specific physiological-anatomical point of view promoted by Haberlandt thrived among the circle of students, assistants, and researchers associated with the Botanical Institute of the University of Berlin from the time of Schwendener's arrival in 1878 to the early 1890s. Schwendener remained at the institute until 1910, but the last two decades of his tenure there were not marked by the energy, enthusiasm, and flurry of research activity that characterized the first twelve years or so. During that initial period he succeeded in encouraging a number of young botanists to explore the next logical stage in the development of physiological plant anatomy: the investigation of the relationship between structure and function in plants living under their natural conditions. Such work was initially restricted to various regions of Germany, but eventually many of Schwendener's students were given opportunities to take their research abroad. The recent unification of Germany and the interest in colonial expansion created favorable conditions under which botanists could secure the means to travel to exotic environments. In the early 1880s Haberlandt himself had such an opportunity and returned with a glowing report of the advantages to be gained from studying plants living under conditions unfamiliar to the native European. By that time, interest in plant adaptation had surfaced on many fronts, and Schwendener's students no longer felt as though they were fighting a solitary battle against an obstinate botanical tradition. Nevertheless, the development in Berlin of an active center for the study of physiological plant anatomy contributed significantly to the intellectual climate that helped bring about that change.

SCHWENDENER IN BERLIN

When Schwendener moved to Tübingen in 1877, he found conditions there little better than in Basel. The botanical laboratories were in a state of disrepair, he had no assistants, and there were many undergraduates in need of basic instruction. Wilhelm Hofmeister, Schwendener's immediate predecessor, had occupied the chair in botany in Tübingen during the last five years of his life. The self-taught botanist whose investigations into the life histories of the lower plants had permanently altered the foundations of his science a generation before was faced during his years at Tübingen with a succession of personal tragedies that strongly affected his work and his health.[1] As a result, Schwendener inherited something less than a vital, active research facility. Haberlandt's recollection of the Tübingen institute was not a particularly cheery one:

> The botanical institute where I immediately began my work was in a fairly desolate state. Schwendener, preoccupied with his investigation of leaf position, had not yet found time to put the institute in order and bring things somewhat up to date. The assortment of available microscopes presented difficulties. We finally determined that an old Plössl instrument – one that dated from Mohl's time [Hugo von Mohl, Hofmeister's predecessor in Tübingen] – was adequate for our purposes, and we set it up for the probationers in a large workroom. I was the only advanced worker there, and twice each week a beginner appeared to learn microscopy. There was no assistant, since on Hofmeister's death his assistant . . . disappeared from the institute.[2]

Schwendener quickly realized that this was not the environment in which to set up a flourishing research program. When he was offered the chair at Berlin only a few months after his arrival in Tübingen, he discussed the matter at length with Haberlandt. Schwendener had a strong distaste for urban life. He knew that he would feel more comfortable in a small city like Basel or Tübingen. As Haberlandt reported it, he was also unconcerned with the prestige that the new position would bring. However, since he could hardly pass up the opportunity to be placed in charge of a first-rate botanical research facility, there was no choice but to accept the offer.[3] He moved to Berlin in the fall of 1878.

The University of Berlin was the largest university in Germany, with the largest budget. Enrollment was still small by present standards; there were 3,608 students there in 1880, but this was three times the number at Tübingen. The difference between those two institutions is more striking when the enrollment figures are broken down by faculty. In 1880 there were 1,621 students in the philosophical faculty in Berlin as compared with 318 in Tübingen; within that faculty, there were 509 in science and mathematics in Berlin as compared with 51 in Tübingen. In addition to the new, well-equipped botanical institute, the University of Berlin also included among its facilities the largest botanical garden and museum in Germany. As a point of comparison, the total expenditure for the botanical institute, garden, and museum in Berlin for the 1892–3 academic year was 137,175 marks, whereas at Tübingen the botanical facilities had a total expenditure that year of 19,040 marks, a figure that was not particularly low among Germany universities at that time.[4]

Outside the university were technical schools and institutes always in need of botanical assistants and instructors with university training. As an added attraction, retired Jena professor Nathaniel Pringsheim maintained a private botanical laboratory at his home in Berlin. Pringsheim, who was living on a substantial inheritance, took an active interest in the botanical program at the university and often employed university students as his assistants for a year or two after they completed their formal training. Soon after Schwendener arrived, Pringsheim persuaded Berlin botanists at the university, the botanical gardens, and the agricultural college that there was a need for a national organization. In 1882 he became the first president of the product of his efforts, the Deutsche Botanische Gesellschaft, the first nationwide botanical organization for the unified German state and perhaps the first scientific organization to encompass the entire German-speaking realm.[5] The Deutsche Botanische Gesellschaft quickly became the central organ of the German botanical community, holding monthly meetings and publishing its journal, *Berichte der Deutschen Botanischen Gesellschaft,* which soon rivaled in importance *Botanische Zeitung* and *Flora,* the two most widely circulated German botanical periodicals in the early 1880s. One could hardly imagine more advantageous conditions under which to initiate a botanical research program.

Schwendener's time had come and he made the most of it, largely through his skill in directing students in individual research projects. As for his own work, this was now limited to theoretical writing; he discontinued active laboratory research altogether. Haberlandt was dismayed, in fact, that he never saw Schwendener working at his microscope either in Tübingen or in Berlin.[6] The testimony of his Berlin students, however, indicated that he was an effective and inspiring teacher who devoted all his energies to that role. In his obituary notice for his former teacher, Albrecht Zimmermann wrote that Schwendener took his teaching responsibilities, especially the training of doctoral candidates, so seriously that he left little time for his own research.[7] He was quite formal and demanding in the classroom and laboratory. Like his own mentor, Carl Nägeli, he emphasized an inductive approach to plant study, discouraging students from consulting the available literature until they had made extensive firsthand investigations on their own. Alexander Tschirch, one of his first students in Berlin, offered the following insight into Schwendener's pedagogical techniques:

> Strongly "philologically disposed" . . . I was drawn irresistibly to the literature, and therefore, despite all dissuasion, I began to study it. However, every time "Master Simon," as we called Schwendener in the laboratory, found me over a book instead of at the microscope, he came over to me uttering the condemning words: "I see that you are engaged in literature work." Yet if he found me at the microscope, he immediately sat at the

> instrument, looked at my preparations, discussed the meaning of the observations, and often remained an hour or longer at my side, lecturing untiringly and, if necessary, taking a hand in the work itself. At that time, in doubtful cases he made many control preparations for me. From him I learned to work carefully and to interpret my observations. He himself was very cautious in his interpretations. All in all, I learned how able a preparator and how critical a judge he was.[8]

This scrupulous observer and critical judge clearly commanded the respect of his students. Yet it was the nature of the research projects that he directed, and not Schwendener's cautious, inductive approach to botany, that gave the Berlin program its unique character and its cohesiveness.

Schwendener's students felt that they were taking part in the creation of an exciting new direction in botanical research; they saw applications for physiological plant anatomy in every branch of botany. Furthermore, the opposition already expressed toward Haberlandt's work gave the program the added sense of excitement and camaraderie that come from banding together in defense against a common enemy. Much in the manner of an old soldier recalling his war years, Tschirch described the mood among Schwendener's students in the early 1880s:

> [Schwendener] was to found an anatomical-physiological school. Haberlandt's *Entwickelungsgeschichte des mechanischen Gewebesystems* appeared in 1879, and already the opposition had come forward. Offensives rained from the school of de Bary, who advocated pure anatomy, and the school of Sachs, who advocated pure physiology. It was important, therefore, to find comrades-in-arms and to further improve upon and fortify the new viewpoint, whose outlines had already been drawn. We were these comrades-in-arms: Haberlandt, who at that time no longer worked with Schwendener, Krabbe, who died young, Ambronn, who later went over to physics, Volkens, Westermaier, Zimmermann, Block, Fünfstück, Reinhardt, Potonié – the trumpeter for the group, as Schwendener called him, since he defended the viewpoint in the press – and myself.[9]

Although not all the members of Schwendener's circle viewed their experience in military terms, they remembered the early days of the Berlin Botanical Institute as an intellectually exciting period in their lives. Otto Reinhardt, who arrived in Berlin in 1881, conveys that feeling in the following passage:

> An exceedingly stimulating scientific life prevailed in our institute at that time – not only in the institute itself, in connection with the instruction of Schwendener, whom we quickly named "The Master," but also in the informal gatherings in a nearby tavern . . . where we discussed the then current questions in the forefront of our science, often with a resulting lively interchange of ideas.[10]

With just a bit of exaggeration, we can say that in only a few short years Simon Schwendener had been transformed, by circumstances and personal choice, from the solitary botany professor at a small Swiss university to Master Simon, the inspirational leader of a group of botanical "comrades-in-arms," who fought scientific battles by day and pondered the pressing philosophical questions of their discipline, over glasses of beer, by night.

SCHWENDENER'S STUDENTS

The Berlin Botanical Institute attracted students with a wide variety of backgrounds, interests, abilities, and goals, and Schwendener managed to accommodate this diversity without sacrificing the identity of his program. The subjects of the projects pursued by students, assistants, and postdoctoral researchers during the early years of Schwendener's botanical institute indicate the diversity of topics included within the province of physiological plant anatomy. Much of the early work reflected Schwendener's interest in mechanical problems as they relate to plant structure. Albrecht Zimmermann (1860–1931), who followed Schwendener from Tübingen to Berlin, completed a dissertation on the mechanical devices that facilitate the dissemination of seeds and fruits; Hermann Ambronn (1856–1927) investigated the mechanics of the coiling process in climbing plants; and Gustav Krabbe (1855–95) studied the effects of the force of weight on the growth of thickened rings in wood tissue.[11] For Schwendener, interest in such mechanical problems never waned; throughout his career at Berlin, he continued to assign students topics relating to mechanical aspects of spiral phyllotaxy. Others pursued what might be called more conventional topics in physiological anatomy. Alexander Tschirch and Georg Volkens (to be discussed in the following section) investigated problems dealing with the relationships between particular plant structures and the moisture and temperature conditions of the natural environment. Max Westermaier (1852–1903), Schwendener's assistant for many years, studied the relationship between structure and function in epidermal tissue and the physiological significance of tannin; and Nordal Wille (1858–1924), a Norwegian botanist who studied briefly under Schwendener, wrote a treatise on the physiological anatomy of the algae.[12] Emil Heinricher (1856–1934) went to Berlin for further study in the early 1880s after receiving his doctorate under Hubert Leitgeb at Graz. The immediate result of his association with Schwendener was a work on the relationship between environmental conditions at various sites throughout Germany and the orientation of the internal structures of leaves.[13]

This early group also included, among others, Moritz Fünfstück (1856–1925), Henry Potonié (1857–1913), and Otto Reinhardt (1854–1935), each of whom was swept up in the physiological anatomy movement at Berlin in the 1880s. The pages of *Botanisches Centralblatt, Botanischer Jahresbericht, Berichte der Deutsche Botanischen Gesellschaft,* and even de Bary's *Botanische Zeitung* were kept filled with notices and reviews of work issuing forth

Gottlieb Haberlandt (left) and Simon Schwendener (seated), along with Carl Correns, in 1905. Correns studied briefly with Schwendener in Berlin. Reproduced by permission of Springer, Heidelberg.

from Schwendener's institute. Much of the publicity was the result of the efforts of Zimmermann and Potonié, who wrote most of the notices themselves. In addition, Potonié published many informal articles by members of Schwendener's circle in his *Naturwissenschaftliche Wochenschrift*, a Berlin-based popular science weekly.[14]

Few of Schwendener's students went on to establish themselves in botany professorships as specialists in physiological anatomy. Many changed fields for one reason or another. Ambronn was so taken up by mechanical problems that he left botany for physics, Tschirch returned to pharmacology after a brief flirtation with physiological plant anatomy, Volkens eventually found a position serving the botanical interests of the new German colonies, and Potonié turned to paleobotany. Since the German university system was not adding new chairs in botany in the 1880s and 1890s, a professor could hardly hope to place even his best students in secure university appointments. A wide range of employment opportunities for trained botanists existed outside the universities, although such opportunities were seldom the priority of those trained within the rigorous academic setting of a university program. Also, since the Austrian universities were not included within the German system, it was still possible for Austrians who received training in Germany, such as Haberlandt, to find positions in one of their native universities, where fewer candidates were competing for available chairs. Two examples should help to indicate the range of career paths chosen by Schwendener's students.

After completing his doctorate in 1881, Albrecht Zimmermann assisted both Schwendener at the botanical institute and plant physiologist Leopold Kny at the Agricultural College in Berlin until 1885. There followed a period of ten years in which he served as a *Privatdozent* in Leipzig, in Tübingen, and in Berlin again under Schwendener while searching in vain for a permanent academic appointment. In 1896 he finally gave up the search and accepted a position at the Coffee Research Station in Java, then associated with the Buitenzorg Botanical Garden. After nearly six years in Java, Zimmermann found a research position at the new Biological-Agricultural Institute in Amani in German East Africa, one of the many research stations established in the new German protectorates. He remained there for eighteen years, serving as director of the institute from 1911 to 1920. He had made minor contributions to physiological plant anatomy during the 1880s and early 1890s, but his work in the colonies forced him to redirect his research interests. After 1896 his publications concerned mainly the cultivation and diseases of economically valuable exotic plants. Although some aspects of his research required him to consider adaptive phenomena, he had to restrict his work to the purely practical side of this problem as it applied to cultivated plants.[15]

Zimmermann's case was extreme. Most of Schwendener's students found employment closer to home. When Emil Heinricher returned to Austria in 1884, he taught at the technical college at Graz while assisting Haberlandt at the university. In 1889 he accepted an appointment at the University of Innsbruck, and he was named to the chair in botany there in 1891. He incorporated physiological plant anatomy into many phases of his research. At Innsbruck he immediately worked to establish a botanical institute and to rearrange the plants in the botanical garden into "biological groups," that is, groups arranged by habitat rather than taxonomic affinities, a project begun earlier by

Anton Kerner. Although he produced no single large work of any significance, Heinricher authored numerous small studies concerning various aspects of physiology, experimental morphology, physiological anatomy, and taxonomy. His major emphasis was on the biology of parasitic plants, a subject for which he produced no fewer than seventy-nine separate monographs during his years at Innsbruck. His work, as well as that of his student Adolf Wagner, revealed the continuing influence of Schwendener and Haberlandt.[16]

Despite the variety of career trajectories followed by the many botanists who passed through Schwendener's institute in Berlin, the research program there remained healthy and vigorous during the 1880s. One new area that Schwendener began to explore through his students was the relationship between anatomical features of a group of plants and particular sets of environmental conditions. This kind of study represented a break from the total laboratory orientation of Schwendener's earlier research. Although such investigations required extensive use of laboratory techniques, they also required a thorough knowledge of the physical conditions of the natural habitats of the plants under study. The work of Tschirch and Volkens best exemplifies this new direction in Schwendener's interests.

PHYSIOLOGICAL ANATOMY AND THE NATURAL HABITAT: THE EARLY WORK OF TSCHIRCH AND VOLKENS

The new environmental emphasis of Schwendener's research program in Berlin was not the result of a belated conversion to natural selection theory. Darwinism was certainly a motivating factor for his students, but for Schwendener himself the need to study the natural environment of the plant was a logical outgrowth of his longtime interest in the interrelationship between structure and function. In addition, having been relocated from a provincial university town to the nation's capital, still buzzing with the excitement of the recent unification wars and now dreaming of colonial expansion, Schwendener could not avoid being caught up in the adventurous national mood that sought to send out tendrils, scientific and otherwise, from the motherland into every region of the globe. The examination of the relationship between the structural and physiological details of plants and their natural environmental conditions was the sort of scientific research program that could lend itself nicely to this mood. It was grounded in sound scientific principles, the best of German laboratory science, and it involved a geographical dimension.

Recent developments in the science of plant geography may also have exerted an influence on Schwendener and provided him with a broadly conceived theoretical model in which to frame particular studies. Floristic plant geography, the study of the distribution patterns of plant taxa, had been receiving considerable attention throughout the nineteenth century, aided and encouraged by extensive explorations by European naturalists into ever more remote areas. Since the appearance of the *Origin of Species,* floristic geographical studies had also been linked to efforts to reconstruct plant phylogenies. However, another branch of plant geography developed alongside floristic studies and eventually found its center in Germany. This was the study of the relationship between climate, or environment in general, and *types* of vegetation, rather than taxonomic groups. In the early part of the century, Alexander von Humboldt contrived a plan to classify plants according to their physiognomic form and to associate each of these forms with a particular set of climatic conditions; he subsequently divided the globe into a series of vegetation zones based upon this scheme.[17] Other botanists continued various aspects of this work, but Göttingen professor August Grisebach (1814–79) elaborated considerably upon Humboldt's series of physiognomic plant forms and his scheme of vegetation zones. In 1872, just a few years before Schwendener came to Berlin, Grisebach had consolidated his life's work in *Die Vegetation der Erde,* a comprehensive two-volume treatise on plant geography based upon the physiognomic tradition. In the years following the publication of that work, a number of botanists in Germany and elsewhere turned their attention to studies involving the relationship between vegetation and environment. In 1875, as a reflection of the increasing interest in this subject area, the annual literature review *Botanischer Jahresbericht* began to include separate categories listing works concerned with the relationship between vegetation and substrate, habitat, and temperature, respectively.[18] A botanist whose major interest was the interpretation of plant structure in terms of function could not fail to see the connection between his own work and this growing tradition in plant geography; and Schwendener quite deliberately exploited this connection with two of his first students in Berlin.

Alexander Tschirch (1856–1939)[19] was born in Guben, to the east and south of Berlin, near the Polish border. His father was a pharmacist and a deacon at the local church. Following his *Gymnasium* education, Tschirch underwent a three-year apprenticeship in pharmacy, in preparation, so his family hoped, for a career in chemistry. With that goal in mind he entered the University of Berlin in 1878, the year that Schwendener arrived. Saturated with the practical training of a pharmacist, and harboring secret aspirations to become a poet or an actor, he quickly fell into the spirit of a university education. In addition to attending courses in pharmacology and chemistry, he took advantage of the distinguished Berlin faculty in other fields. He heard lectures from, among others, the physiologist Emil Du Bois-Reymond, the pathologist Rudolf Vichow, the physicist Hermann Helmholtz, and the historian Heinrich von Treitschke. He also heard the Berlin botanists, August Eichler and Simon Schwendener, and he decided to pursue his doctorate in botany, a field with close ties to pharmacology in the nineteenth century.

When he came to Schwendener as a doctoral candidate, the latter assigned him the task of determining whether anatomical differences in plants reflect the different vegetation zones outlined by Grisebach. Schwendener was excited by the prospect of applying physiological anatomy to plant geography, but he was apparently unaware of the scale of the problem. It did not take Tschirch long to realize the immensity of his assigned task. With Schwendener's somewhat reluctant approval, he reduced the topic to a consideration of the mechanism of ventilation in plants, the stomatal apparatus. In particular, he compared the anatomical structure and arrangement of the stomata (openings in the plant epidermis regulated by the contraction of pairs of guard cells) in plants found in wet and dry habitats. With a bit more persuasion, Tschirch convinced Schwendener that he should not even present his study within the context of Grisebach's vegetation zones. Although he found differences in the location of the stomata between plants in wet and dry habitats, he did not discover any simple correlation between structure and habitat that justified linking these results directly with Grisebach's general scheme.[20]

Recalling the criticism that Haberlandt's *Entwickelungsgeschichte* had received, Schwendener advised Tschirch to publish in a journal in which de Bary and his students were not involved. With a little searching Tschirch turned up *Linnaea,* an old Berlin-based botanical periodical that was encountering serious financial difficulties. The journal's aging editor was more than willing to take on the task, and the result was an adequate low-budget reproduction of the dissertation.[21] Two years later, when Tschirch met de Bary for the first time at a scientific meeting in Freiburg, the latter, never at a loss for wit, remarked: "I had imagined you entirely differently. You are still quite young and adroit, but judging by the content of your dissertation one should think that you were already quite old, and judging by the terrible paper on which your work is printed, that you must wear a shabby coat."[22] De Bary's comment on Tschirch's apparent age was a reference to the teleological character of his work. For de Bary, studies of plant adaptation belonged with the speculative science that had preoccupied German botanists earlier in the century; to interpret plant structure in terms of function seemed to him an approach better suited to *Naturphilosophie* than to scientific botany. However, de Bary had misunderstood the significance of Tschirch's dissertation. Earlier work in plant geography had developed quite independently of plant physiology. Although the association of particular plants, or plant forms, with particular habitats was common knowledge among naturalists, little effort had been made to link the plant with the habitat physiologically. This was due in part to the background and point of view of early plant geographers – these were often men who had traveled extensively, viewed vegetation in terms of large-scale patterns, and had some training in taxonomy but not physiology – and in part to the rudimentary development of plant physiology before the 1860s. Tschirch's work on stomata represented a new trend in which laboratory-trained botanists applied their knowledge of plant anatomy and physiology to particular aspects of the study of plant distribution. Although still relatively new in 1881, this approach would gain popularity throughout the 1880s and 1890s and would eventually develop into what came to be known, largely through the efforts of A. F. W. Schimper (see Chapter 7), as "physiological plant geography."

Tschirch had numerous copies of the dissertation printed and sent them to botanists throughout Europe, in exchange for which he received generally favorable correspondence. Like Haberlandt, he sent a copy to Darwin, who sent back a cordial reply, complete with a picture of himself.[23] Tschirch's high regard for Darwin is evident in this excerpt from a later paper entitled "The Significance of Leaves in Nature's Economy":

> No scientific theory has had a more beneficial influence on biological research than that of Charles Darwin, which, as long as men practice science, will be viewed as one of the shiniest stars in the heaven of exact research. The time is gone when empty descriptions alone constitute the sciences, when morphological subtleties excite the spirit. Now we have advanced to the question "Why?" Not that we no longer consider questions of anatomy, morphology, and developmental history, or even that we underrate their significance within the framework of science – not at all. However, we have found that the time has come, before something new is provided, to utilize and work up, in a biological-physiological sense, the vast amount of information that has accumulated – not to pluck, dry out, and describe all the fruits along the way without any fixed plan, but to put questions before nature and determine, in an experimental manner, how we may discover clues as to nature's mysterious rules.[24]

Tschirch's appeal for synthesis and causal explanation resembles Haberlandt's earlier decision to assemble the disconnected fragments of physiological plant anatomy into a single textbook. Both botanists insisted that Darwin's evolution theory has given us the proper framework in which to ask the question "Why?"; it not only justifies, it necessitates the shift from description to explanation in the biological sciences. This message would be repeated, in one form or another, by all of the botanists under consideration here.

Tschirch himself did not exploit the new viewpoint to any great extent. After completing his dissertation, he worked for a short while as an assistant to Nathaniel Pringsheim at his private laboratory, and he later assisted A. B. Frank at the Plant Physiology Institute of the Agricultural College in Berlin. Tschirch taught botany and pharmacology at the college and at the University of Berlin throughout the 1880s. He was also a founding member, as were many of Schwendener's students, of the Deutsche Botanische Gesellschaft in 1882, and he served as one of the secretaries of that organization for the first eight years of its existence. Then in 1890 he took a position in pharmacology at the University of Bern, thus returning to his original field. He remained at the Swiss university for the rest of his professional career. Tschirch published numerous botanical papers during the 1880s, most of them dealing with chemical-physiological studies of chlorophyll. Some of these papers were written from the point of view of physiological plant anatomy; others reflected the more traditional views of Pringsheim and Frank. After 1890 Tschirch's publications, naturally enough, turned almost exclusively to pharmacology. However, Tschirch did have an opportunity to share in one of the experiences that provided European botanists with an important source of stimulation for their newfound interest in plant adaptation when in the winter of 1888–9 he made an excursion to the British and Dutch Indies. The trip included a visit to the laboratory of Dutch botanist Melchior Treub in Buitenzorg, Java, where Tschirch conducted a physiological investigation of the seeds of tropical plants. Struck immediately by the extreme diversity and manifold adaptations of tropical vegetation, which he had the opportunity to observe for the first time in its natural setting, Tschirch, like many of his contemporaries, wrote an enthusiastic report on his return home and encouraged his colleagues to make similar journeys.[25]

For a European botanist trained primarily in the laboratory, firsthand observation of exotic vegetation growing in a totally unfamiliar environment could be a startling and revelatory experience. If the experience came early enough in his career, it could have a profound effect on his future. This was the case with Georg Volkens, one of Tschirch's fellow students in the early years of Schwendener's botanical institute. However, before taking his research program abroad, Volkens first received a thorough grounding in the methods and point of view of physiological anatomy. He was one of the strongest advocates of the Haberlandt–Schwendener approach to botany during the 1880s.

Georg Volkens (1855–1917)[26] was born in Berlin and, despite his extensive travels, remained a Berliner all his life. His father, a master tinsmith from Heide in Holstein, died of cholera the year he was born, and his mother eventually married another master tinsmith. Volkens attended a *Realgymnasium* in Heide and entered the University of Berlin in the mid-1870s. He began his training in botany there under the aged Alexander Braun, who gave all of the major series of lectures at that time and directed the botanical garden. Sensing that botany had gone off in new directions since the days of Braun's prime, Volkens went to Würzburg to study briefly with Sachs, but he found himself under increasing pressure from his stepfather to give up the dream of a scientific career and to prepare for a more sensible career as a high school teacher. Considering the difficulty of obtaining a university chair at that time, perhaps his stepfather was right. However, the deaths of both his stepfather and Alexander Braun provided Volkens with the incentive to follow his own desires. He returned to Berlin to work with Schwendener, taking his doctoral degree in 1882. Shortly thereafter he arranged the first of several scientific expeditions that would take him eventually to Egypt, East Africa,

Java, and the Caroline, Mariana, and Marshall Islands. As a result of his travels, Volkens became an enthusiastic proponent of German colonial development. He never held an academic chair. He worked as an assistant at Schwendener's institute and at the Berlin Botanical Garden and Museum during the 1880s and 1890s, and through the museum he eventually procured a position as Germany's colonial botanist.

Volkens's research proceeded by quantum leaps through a succession of ever more broadly conceived projects. His doctoral dissertation concerned a fairly narrow aspect of physiological plant anatomy, although he did challenge no less a figure than Julius Sachs. Against some opposition, Sachs had proposed that in leaves the flow of water is directed through the imbibing membranes, that is, the individual cell membranes, whereas the leaf vessels serve only as aeration devices, much like the tracheae in insects. In contrast to this view, Volkens's work showed that the vessels play an important role in the water transport system. From the simple observation that water droplets tend to collect on the borders of leaves in the vicinity of the ends of vessels, Volkens carefully traced the flow back into the leaf, noting that water passes through the vessels under all climatic conditions, wet or dry, thus dispelling Sachs's aeration theory. True to the spirit of physiological anatomy, in order to determine precisely the functional significance of the vessels, Volkens examined these structures indoors and outdoors in plants growing under a variety of moisture regimes.[27]

His next undertaking engaged him in the problem that Schwendener had posed for Tschirch regarding the relationship between habitat and anatomical structure. Whereas Tschirch had confined his work to a single anatomical feature, the stoma, Volkens decided to take on a process – transpiration (water loss by evaporation). He compared taxonomically similar plants living in wet and dry habitats and concluded, against the prevailing view, that transpiration is a purely physical process, a consequence of the way plants are made. A number of botanists, most recently Paul Sorauer, a specialist in plant diseases at the Berlin Agricultural College, had argued that transpiration is a purely physiological process regulated by the plant. Sorauer claimed that atmospheric temperature and moisture at best exert only an indirect influence on transpiration. Volkens argued that such a view could not account for the many adaptations against desiccation displayed by plants found in dry habitats, and he proceeded to describe these adaptations, species by species, pointing out the anatomical differences between similar species found in different climatic zones.[28]

This work on transpiration set the stage for Volkens's next and still more ambitious project: an anatomical-physiological study of desert vegetation. Volkens conceived of this study as no less than a model for the revamping of plant taxonomy along Darwinian lines, making adaptation to different environments the basis for separating closely related taxa. Since central Europe is not known for its deserts, carrying out this project required a lengthy excursion to Egypt, for which Volkens had to obtain the assistance of Germany's most distinguished scientific academy. This was one of the first of many excursions outside the borders of the Reich by the new breed of German botanists. For Volkens, far beyond his original intentions, it provided an experience that shaped a scientific career in the service of Germany's new colonies.

5

Physiological anatomy beyond the Reich

The botanical excursion abroad had been a major facet of the science of botany since the mid-eighteenth century, when Linnaeus's students, emulating their mentor's profitable journey to Lapland, set off to scour the globe in search of as yet undiscovered species. For over a century European botanists had swelled the collections of museums and universities, filled the pages of journals with reports of their new discoveries, and produced monograph after monograph and flora after flora, delineating and describing the species of a given family or genus or the complete catalog of the plant inhabitants in a given locale. This phase of botany continued to be exploited with profit throughout the nineteenth century as imperial ventures brought Europeans into closer contact with vast regions not yet fully explored. As one might expect, however, the changes that took place in botanical science since the mid-nineteenth century led to a different kind of botanical excursion. Schleiden had decried the collecting expedition early on, but a new breed of botanist was finding ample justification for carrying on Schleiden's program of scientific botany beyond the confines of the laboratory.

THE ADAPTATIONS OF DESERT PLANTS

In 1884, while working as Schwendener's assistant in Berlin, Volkens contrived a plan to satisfy a lifelong desire to travel to Africa while also carrying his research to its next stage. Schwendener had made a casual remark that the relationship between habitat and plant structure can best be studied in a region with an extreme climate. Volkens immediately suggested to Schwendener that he try to persuade the Berlin Academy of Sciences to finance an excursion to the Egyptian-Arabic desert. The Berlin Academy had been sponsoring scientific expeditions to Africa, South America, and the islands of the Pacific since the 1860s. These expeditions were linked to Germany's colonial interests, which before the 1880s centered on private economic ventures. After 1884, however, when Bismarck made his formal, if reluctant, appeal for the acquisition of colonies, these expeditions took on a greater political significance.[1] Volkens decided to take advantage of this situation, and Schwendener agreed. With a minimum of effort he persuaded the Berlin Academy to provide Volkens with 5,000 marks for the journey.[2]

The excursion lasted ten months, from the fall of 1884 to the summer of 1885, and resulted in a significant contribution to the study of plant adaptations. In Volkens's own immodest account:

> The fruit of the excursion itself was the work, illustrated with 18 plates, *Die Flora der ägyptisch-arabischen Wüste, auf Grundlage anatomisch-physiologischer Forschungen dargestellt.* This book was a success; it called forth an entire literature and thereby helped found and develop a special discipline of botany, the ecology of plants.[3]

Volkens perhaps exaggerated the role of his book in the development of plant ecology, but his journey to Egypt marked an important shift in point of view for botanical excursions. In the past, botanists who traveled to distant regions for scientific purposes did so mainly to add to their collections, discover new species, or study patterns of distribution. Volkens had none of these goals in mind; he went to Egypt to examine desert plants from a physiological-anatomical point of view. His *Flora* was not a traditional flora at all but an account of structural–functional relationships in selected desert species.

In a preliminary sketch for the book published several months after his return to Berlin, Volkens explained his approach to his subject. Like Haberlandt and Tschirch, he made an appeal for a new methodology and stressed the need to understand the functional significance of plant structures:

> Every plant that confronts us in nature should be examined not just with respect to its kinship to other vegetation. Apart from floral characteristics, which often reflect only a particular relationship to the insect world, it is important also to probe as far as possible into vegetative structure, particularly in its connection to external influences. We know that light and heat impress their stamp on the form and structure of assimilation and transpiration organs, just as the physical and chemical nature of the soil does so for the absorptive organs. Every plant appears to us, therefore, as a product of its surroundings; and among all the characteristics which it "has inherited from its ancestors," we are interested in those among this splendid multitude in which the uniqueness of the habitat is expressed.[4]

Volkens repeated this passage verbatim in the book, which appeared a year later. In emphasizing the need to examine vegetative as well as floral structures, his intention was not to replace traditional taxonomy but to supplement it with studies of the relationships among form, function, and habitat. Volkens believed that studies of plant adaptation, developed alongside taxonomic studies, could lead to important insights into the evolutionary history of the plant groups in question:

> The synthesis [in traditional systematics] has been completed. It is important now to turn away at once from the view of kinship and to ask instead: Which

> are the factors that shape the manifold plant forms? What causes the breakup of an originally monotypic species into numerous species? How have the different characteristics of these species originated in the struggle for existence, and what is their relationship to present or past living conditions? A completely satisfactory answer will never be given, just as a complete understanding of heredity is unlikely to be attained. The result will be a modest one; yet we will find it satisfactory insofar as it contains not bare facts but explanations, or at least insofar as it creates the means for such.[5]

Here was the now familiar appeal for causal explanation, justified in Darwinian terms. Where Haberlandt and Tschirch called for a reform of plant anatomy and morphology, Volkens called for a reform of plant systematics, but the message was the same: The old methodology emphasized description and approached scientific problems piecemeal; the new methodology would emphasize explanation and would integrate studies of structure and function with a knowledge of environmental conditions.

In the introduction to *Die Flora der ägyptisch-arabischen Wüste,* Volkens outlined his plan for a reform of plant systematics. "The only method that I consider possible to bring us closer to our aspired goal," he stated, ". . . is the anatomical-physiological method," and he offered an admittedly ideal example of what he had in mind. In order to work up a monograph of a genus, Volkens explained, a future botanist would first examine and compare the anatomical and morphological features of roots, stems, leaves, flowers, and fruits in all native species belonging to that genus, noting that some of the differences in these features can be attributed to different biological requirements reflecting the various modes of life and habitats represented by the different species. He would then make an extensive examination of available taxonomic records, searching for every representative of the genus to be found anywhere in the world. Gradually, this researcher would compile a list of characters shared by all members of the genus and a second list of characters that vary considerably from species to species or are found in some species and not in others. This second list would be the more valuable of the two. With his extensive knowledge of plant adaptation in its anatomical and physiological details, he would understand that the variable characters represent the intricate relationships between individual species and their respective habitats. He would then determine the environmental factors that caused the first representative of the genus to split up into several species and, in this way, he would eventually reconstruct the entire phylogenetic history of the genus.[6]

The future would have to wait. Volkens knew that the means were not at hand to work up a taxonomic monograph in this way. Still he wanted to offer his study of desert plants as an example, however sketchy and incomplete, of this new direction in botanical investigation. He devoted the first half of his book to a discussion of the physiological needs of desert life as they manifest themselves in plant tissue systems. After briefly describing the physical conditions of the northern Sahara and the general characteristics of the vegetation, he treated at length the processes of absorption, transpiration, and photosynthesis as they apply to desert plants. He followed this discussion with very brief sections on the mechanical system, the vascular system, and flowers and fruits. In the second half of the book, the only part that vaguely resembles a traditional flora, he offered a summary discussion, family by family, of adaptive features found in particular taxonomic groups in the Egyptian-Arabic desert. The book was more a synthesis than an original botanical treatise. Volkens made liberal use of the recent work of plant physiologists Friedrich Kohl and Hubert Leitgeb on transpiration and the physiology of stomata, respectively, and the work of Gottlieb Haberlandt and Ernst Stahl on photosynthesis, and he amplified this material with specific examples from his own research on desert plants.[7] The original feature of Volkens's work was his integration of this material in a study dealing with plants living in one region under a particular set of environmental conditions.

Two examples will illustrate his approach. The first has to do with the problem of transpiration. Not surprisingly, Volkens devoted the most space in the book to the chapter on transpiration. This was the process he had studied most extensively prior to his trip to Egypt; and for a committed Darwinist, it made sense to focus one's attention on adaptations against water loss, the most serious survival problem encountered by desert vegetation. As in his earlier article on this subject, Volkens challenged the prevailing view that transpiration is a physiological process. In this instance he focused his critique on Friedrich Kohl's recent book, *Die Transpiration der Pflanze*. After praising Kohl's thoroughness and accuracy, Volkens offered a point-for-point rebuttal of Kohl's case for claiming that transpiration is a physiological process. Kohl's argument was based upon three observations: (1) transpiration is greater in dead leaves than in living ones, (2) transpiration is greater in light than in darkness, and (3) the transpiration rate per unit of leaf surface area increases after part of the leaf is removed. Whereas Kohl considered these observations as evidence for the active role of the living plant in the regulation of transpiration. Volkens offered simple physical explanations for each of them. Regarding the first, Volkens argued that plant tissue is altered upon death and that plant cells lose their capacity to retain water. This excess free water evaporates, and one observes an increase in transpiration. Regarding the second observation, Volkens claimed that there is no direct connection between light and the increased transpiration rate. Kohl attributed the increase to heat generated by the chemical activity of photosynthesis, but Volkens argued that this hardly serves as evidence that transpiration is fundamentally physiological in nature. Volkens regarded the third observation as Kohl's most substantial piece of evidence. Studies done on *Helianthus* and *Nicotiana* (sunflower and tobacco) had shown that the transpiration rate per unit of leaf surface definitely increases when part of a leaf is removed. Here Volkens argued that the removal of part of a leaf upsets the balance between absorption and transpiration. Although water is continually drawn upward into the leaf, the path of departure (i.e., evaporation from the leaf surface) has been partially blocked by the reduction in the surface area of the leaf. This blockage leads to an increase in turgor pressure, which, in turn, causes the stomata to open wider, with the resulting increase in the transpiration rate. The simple mechanical response of the stomata to increased turgor pressure, Volkens contended, could hardly count as strong evidence that transpiration is a physiological process.[8]

Why did Volkens take the trouble to refute Kohl's case for the physiological nature of transpiration? He was making his own case for the close correlation between the harsh desert environment and the anatomical details of desert plants. He believed that only by viewing transpiration as a physical process comparable to evaporation from a free water surface can one account for the abundant and complex plant adaptations against the dangers of water loss:

> Only by admitting the external agents, atmospheric temperature and moisture, as the exclusive causal factors [in transpiration] is one allowed to infer the intensity of transpiration from the perceptible structures of a plant. Only in that case may one allow oneself a judgment, such as: this plant is better adapted to the dryness of the climate than that one because of such and such anatomical characteristics.[9]

Volkens viewed the subject of adaptation from a Darwinian perspective. He believed that many anatomical features of desert plants could not be understood except as adaptations to the extreme dryness of the climate; and as a strict Darwinian and a proponent of physiological plant anatomy, he was convinced that every detail of plant structure held a significance in the life of the plant. Every detail, as it were, had to have a purpose. If Kohl was correct in claiming that plants could somehow regulate transpiration from within, then what purpose could there possibly be for the many structural peculiarities of desert plants – the diminished surface area due to reduced or absent leaves, the extra-thick waxy cuticles, the oily surface secretions, the use of the epidermis as well as inner tissue for water storage, the special means of keeping stomata open, and so on? Volkens went on to discuss each of these structural peculiarities in turn, providing abundant illustrations from the desert flora of the northern Sahara.

The second example of Volkens's approach concerns his discussion of the role of the intercellular spaces in the photosynthetic tissue of leaves. These spaces are created by the irregular shapes of the cells that make up the spongy parenchyma, the tissue layer adjacent to the chlorophyll-rich cells of the pallisade parenchyma, where most of the photosynthetic activity takes place. The pallisade cells are regular, elongate, and densely packed, with very thin spaces between them that extend into the spongy layer. The waxy cuticle of the leaf epidermis prevents the outside air from coming into contact with

Georg Volkens. Courtesy of the Hunt Institute for Botanical Documentation, Carnegie Mellon University, Pittsburgh, PA. Reproduced by permission of Borntraeger, Stuttgart.

pallisade cells, but the openings in the epidermis (the stomata) permit the air to reach the pallisade cells through the intercellular spaces in the spongy layer. These spaces vary considerably from species to species; and there was a debate at the time of Volkens's writing as to whether this variation reflects different adaptive responses to transpiration or to the rate of photosynthesis. Haberlandt, among others, took the view that enlarged or constricted intercellular spaces serve as means to either facilitate or hinder transpiration. Ernst Stahl maintained that the size of the intercellular spaces is a reflection of the photosynthetic requirements of the plant in accordance with its given light conditions. Believing that his study of plants living in the harsh desert environment offered an excellent opportunity to resolve this debate, Volkens used leaf cross sections to calculate the volume of the intercellular spaces in selected desert species, and he compared these results with photosynthetic rates, light conditions, and specific knowledge of the habitats of the respective plants. He had to side with Stahl. He found a close correlation between the volume of the intercellular spaces and photosynthetic activity, yet almost no correlation between the volume of the spaces and transpiration. Many species equipped with elaborate means of protection against dryness nevertheless had the same loose arrangement of photosynthetic tissue, that is, large intercellular spaces, that is found in species living in humid zones.[10]

In neither of these examples did Volkens doubt for a moment that the particular structures or patterns of organization he was examining had adaptive significance. This view was consistent with his strict selectionist interpretation of evolution and, of course, with physiological plant anatomy. It was this confidence that all anatomical details have functional significance that led Tschirch, Volkens, and their contemporaries to focus their work on the environment of the plant, to seek correlations between the particulars of plant structure and the particulars of habitat. Since their belief in the close fit between organic structure and environment was based upon Darwinism, they felt assured that their discussions of purpose would not be considered "teleological," except in a causal-mechanistic sense. Volkens criticized Friedrich Kohl for avoiding discussions of adaptation and purpose in his book on transpiration. It is only by making exhaustive inquiries into purpose, Volkens argued, that we can begin to understand the details of organic structure.[11]

Although Volkens's study of the adaptations of desert plants may not have represented a major theoretical breakthrough, it effectively combined the methods of the laboratory botanist with those of the field researcher in a study confined to a particular geographical region; it represented one of the earliest ecological field studies conducted by a laboratory-trained botanist. The book appeared in an oversize, large-print edition with full-page, detailed anatomical illustrations. The reviewer for *Botanische Zeitung,* A. Fischer, offered sarcastic comments regarding the format of the book but otherwise praised its contents. Fischer more than once mentioned that Volkens presented little that was not already included in Haberlandt's *Physiologische Pflanzenanatomie,* but he agreed that the work contained much information that would prove to be of use to plant geographers. He had serious reservations, however, concerning Volkens's conviction that future studies carried out along similar lines would prove valuable in reforming plant systematics and would shed light on evolutionary questions.[12] Absent from this review was the vindictiveness of the earlier attack on Haberlandt's *Entwickelungsgeschichte des mechanisches Gewebesystems.* By 1887 even botanists who may have rejected the physiological-anatomical point of view in their own work did not express their objections, in principle, to this sort of undertaking.

THE COLONIAL BOTANIST

In the time between his return from Egypt and the publication of *Die Flora der ägyptisch-arabischen Wüste,* Volkens obtained the advanced degree required for university teaching (the *venia legendi*) but had searched in vain for an academic appointment, owing, as he explained, to the overabundance of young botany Ph.D.s at that time.[13] When Adolf Engler was appointed director of the Berlin Botanical Garden and Museum, to replace the late August Eichler, Volkens went to work for him as an unpaid volunteer. Engler was soon to become Germany's premier systematist and floristic plant geographer and would develop the Berlin garden and museum into a showcase for the botanical treasures from Germany's new colonial acquisitions. He would also offer continual support and encouragement to Volkens throughout the latter's career, especially with regard to his colonial ventures. During his initial tenure with Engler, however, Volkens discovered that traditional taxonomic work was not to his liking. He produced two taxonomic monographs, not at all conceived on physiological-anatomical principles, and then returned to Schwendener's institute, where he was supported by a small stipend until 1891.

In the winter of 1891–2 he worked out a plan to resume his old line of research in Africa. Having already examined desert plants, Volkens now wished to study the adaptations of plants living on the tops of mountains, another environment subject to the extremes of climate. His first choice was to travel to South America, but he was unable to find the means for such a journey. Realizing that Mount Kilimanjaro, in the newly acquired German protectorate of East Africa (now Tanzania), might serve his purposes equally well, he once again approached Schwendener for assistance. As before, Schwendener used his influence in Berlin to secure Volkens's financial support. He made the necessary arrangements with the Colonial Division of the Foreign Office and also with the Berlin Academy of Sciences, which provided a grant from the Humboldt Foundation. Volkens's expedition was to leave for Kilimanjaro in the fall of 1892, and it was to be joined with that of geologist Carl Lent and a forester named Wiener, sponsored by the Deutsche Kolonialgesellschaft.

Volkens arrived in East Africa in the summer and began making preparations. Unfortunately, he was soon greeted by the news that a local uprising had driven German colonial troops from Kilimanjaro to the coast. The expedition would have to be delayed indefinitely. When conditions again seemed favorable, Volkens, Lent, and Wiener gathered at Tanga, on the coast, in the early winter, only to be told that their expedition would have to be postponed until the following March. Meanwhile, they had plenty of time to become acquainted with the colonial officers, to familiarize themselves with the local customs and the inhabitants, to make short field excursions, and to gather provisions for their longer journey. When they finally made the ascent, Volkens was disappointed; the mountain environment did not lend itself to the kind of physiological–anatomical investigation that he had intended. He had hoped that the anatomical structures of the mountain plants would somehow reflect the effects of height and/or snow cover; instead, due to the intense insolation at the higher elevations, they reflected mainly the effects of extreme dryness, a subject that he had already addressed in his study of desert plants. Turning from his original plan, Volkens began collecting taxonomic, geographic, and ethnographic observations, which he compiled in a book, *Der*

Kilimandscharo, published a few years after his return to Germany, a combination natural history and traveler's journal.[14] His traveling companions did not fare very well. Wiener became seriously ill early on and had to return to Europe; Lent remained in East Africa after Volkens had left and was killed in an ambush. Volkens himself contracted a severe case of malaria that continued to plague him long after his return. Experiencing the exotic was not without its hazards.

Back in Berlin, Volkens remained officially attached to Schwendener's institute for three years, 1894–7, while spending most of his time at the botanical museum. As a result of his African experiences, Volkens had become a spokesman for colonial development. He was one of the founders of the Berlin-Charlottenberg Division of the Deutsche Kolonialgesellschaft, was elected to its board of directors, and gave lectures throughout Germany on the importance of developing the colonies. In 1897 he became a scientific assistant at the Berlin Botanical Museum, and in the following year he was appointed curator in charge of the Botanische Zentralstelle (Central Botanical Bureau), an office created by Engler in 1891, in cooperation with the Colonial Division of the Foreign Office, as a kind of clearinghouse for botanical information relating to the colonies. Earlier that same year (1898), Volkens turned down an academic appointment at Bonn because (according to his own third-person account) "as a Berliner bound up with the culture of that city, and at the same time still living in hope of obtaining an independent position as colonial botanist, the assistant professorship in Bonn . . . was not sufficiently attractive to allow him to become unfaithful to the capital city of the Reich."[15] Having turned down the Bonn position, Volkens abandoned all hope of obtaining a permanent academic appointment in the future; but he took his new role as colonial botanist quite seriously, since he was interested in helping to develop the colonies into productive agricultural regions. He made numerous contributions to the literature on colonial agriculture and occasionally instructed his fellow countrymen in tropical agriculture and horticulture at the Berlin gardens.[16] As the director of the Botanische Zentralstelle, he was also responsible for sorting through thousands of plant specimens sent to Berlin from the various colonial territories and for supplying the experimental gardens in the colonies with living plants from different regions in an effort to test their suitability as crop plants.[17]

No sooner did Volkens assume his responsibilities at the Botanische Zentralstelle when he was asked to accompany the German flagship expedition to the Caroline and Mariana Islands, Pacific islands recently acquired from Spain. He was to be sent as a provisional scientific officer to assess the economic value of the new colonial acquisitions. His ten-month journey, from the early autumn of 1899 to the summer of 1900, took him to the Marshall Islands and North East New Guinea (German protectorates since the 1880s), as well as to the Carolines and Marianas. He spent the bulk of that time, seven months, on the island of Yap in the Carolines and used his observations there as the basis for his report.[18] In 1901–2 Volkens returned to the South Seas to pay a visit to the botanical garden at Buitenzorg, in order to supply the German colonies with useful and ornamental plants. During his stay of over eight months he sent hundreds of seeds, bulbs, tubers, rhizomes, and living plants from Buitenzorg to the various German colonies and also to the Botanische Zentralstelle. Although he was in Buitenzorg in his official capacity as a colonial botanist, Volkens conducted one purely scientific investigation at the botanical laboratory, an inquiry into the causes of leaf fall and leaf renewal in the tropics. This was his last serious attempt at physiological plant anatomy, and his busy schedule and preoccupation with the economic interests of the colonies kept him from publishing the results for a decade.[19]

THE GERMAN COLONIAL ENTERPRISE

Serious German commercial interest in overseas territories dates at least to the 1830s and 1840s, when merchants, largely from Hamburg, first launched trade and light manufacturing ventures on the African coasts and on Pacific islands. This activity increased steadily during the 1850s and 1860s, so that by the mid-1870s German companies had established active trade centers, along with small settlements, on the west and southwest coasts of Africa, in East Africa (in the area that is now Tanzania), and on the island of Samoa. In South America, particularly Brazil, the colonial venture took the form of agricultural settlements. Population pressure and economic setbacks at home led to large-scale emigration of working-class Germans to the Americas in the mid-nineteenth century. In South America, the new settlements became markets for German goods. The German colonists, in turn, controlled local trade and sometimes helped to establish industries in the surrounding regions. Blumenau, where Fritz Müller lived after 1852, was one of several large German "colonies" in southern Brazil, where the cooler climate was more suitable for Europeans. There were German settlements in Chile, Argentina, Uruguay, parts of Central America, and Mexico as well.[20]

The formal establishment of overseas protectorates came about in the 1880s as the result of political pressures at home, diplomatic maneuvers with rival colonial powers, and negotiations with local inhabitants in the colonial regions. In general, the political boundaries of the new protectorates followed closely the spheres of influence of German commercial enterprises. This is not surprising, since the numerous local treaties that defined the territorial limits of German control were often the result of negotiations carried out by German entrepreneurs themselves. Karl Peters, founder of the Deutsche Kolonialgesellschaft, personally negotiated treaties for some 60,000 square miles of African land that was placed under the jurisdiction of the German East Africa Company, of which he was the chief sponsor. In 1885 this region officially became a German protectorate. Through similar guile and tenacity, F. A. E. Lüderitz, a Hamburg tobacco merchant, negotiated treaties with the inhabitants of southwest Africa (in the region that is now Namibia) and, in 1884, succeeded in persuading the Reich to grant official protection to that region. During the same year, protectorates were established in Togo and Cameroon in west-central Africa. In the Pacific, political negotiations centered on protecting the interests of the Hamburg shipping firm of J. C. Godeffroy and Son, which had expanded its original coconut oil trade to include the preparation and export of copra, a much more lucrative coconut product. Initially limited to Samoa, the firm extended its activities to nearby islands and established plantations for growing coffee and cacao as well. Fear of encroachment by rival western powers on the territory controlled by the Godeffroy company and other German firms led Bismarck in 1886 to negotiate a settlement with the British that gave Germany control of the Marshall, Caroline, and Mariana Islands and part of New Guinea. Samoa was eventually divided between Germany and the United States after the two nations narrowly averted warfare over control of the island.[21]

The German presence in South America continued as a series of independent colonization efforts and private commercial ventures. In 1872 the colonists of the large southern Brazilian settlement of Rio Grande do Sul, whose number had reached 60,000, petitioned the German government for formal protection and active encouragement of German commercial interests there through the establishment of an official colonial policy. The Reich granted consular service, but the extent of its political intervention there, as in other South American settlements, remained at the level of diplomacy. Nevertheless, German influence in South America was extensive; German investment and trade persisted at a brisk pace throughout the nineteenth century, and Germans played a prominent role in the affairs of several South American states.[22]

The situation was somewhat different in the new German protectorates, which did not share South America's long history of European domination. Technically, although the Germans called them *Schutzgebiete* in all their official publications, these were not protectorates at all but legitimate (if that term has any meaning in this context) colonies; that is, they were not administered largely by the private companies with interests in the respective regions (Bismarck's original intention) but by a network of officers sent by the German government. Serious economic development did not begin in these regions until after the turn of the century, and even then the entire colonial empire accounted for less than 1 percent of all German overseas trade. There had not been overwhelming support at home for the acquisition of colonies; German financiers were reluctant at first to invest in colonial enterprises, and the government itself invested very little capital in overseas expansion before the early twentieth century.[23] In Africa, the German colonial administration was somewhat haphazard at first and preoccupied with the conquest of inland territories. By the beginning of the First World War, there were still large tracts of colonial lands over which Germany exercised only marginal control.[24] In 1899, at the conclusion of his informative study of the German colonies, geographer Kurt Hassert could only claim: "The significance of our overseas possessions lies, therefore, much more in the future than in the present; and the German colonies must not be valued according to what they cost and what they yield today but according to what they will someday yield."[25]

EARLY SCIENTIFIC INTEREST AND THE ROLE OF TREUB'S LABORATORY

If the economic exploitation of the colonies had to await the future, scientific exploitation had already begun. Among the special interest groups that lobbied in support of colonial expansion in the 1870s and early 1880s, geographical societies, along with academic geographers, figured fairly prominently.[26] The Academy of Sciences of Berlin sponsored several major expeditions to South America, Africa, and Micronesia between 1863 and 1883. These expeditions were largely geographical in character and obtained their financial support primarily from the Alexander von Humboldt Foundation for Scientific Research and Travel. After 1884, when Germany formally acquired a colonial empire, the expeditions became more frequent and longer in duration.[27] Botanists, as well as other scientists, were able to obtain support from the Berlin Academy of Sciences for conducting pure research in regions in or near the new protectorates. The academy, as noted earlier, sponsored Volkens's excursions to Egypt in 1884–5 and to East Africa in 1892–3 and an excursion by A. F. W. Schimper to Ceylon and Java in 1889–90. Its Austrian counterpart, the Vienna Academy of Sciences, sponsored Haberlandt's excursion to Java and Malaysia in 1891–2.

As the itineraries of the previously mentioned journeys indicate, scientific interest extended to regions controlled by other European powers as well as Germany. Before the turn of the century, the African colonies remained inaccessible to all but the more adventurous scientists and explorers, such as Volkens. The more sedate, laboratory-oriented scientists had to settle for travel in regions where the accommodations were more genteel and where facilities existed for serious research. Aside from concern for their personal safety and comfort, the botanists under consideration in this study could hardly have been content to travel in regions accessible only to collectors. For their particular set of scientific interests, Treub's laboratory in Java offered a nearly perfect setting.

Melchior Treub (1851–1910) shared many of the views of his German contemporaries. He received his botanical training in Leiden in the 1870s

Melchior Treub. Courtesy of the Hunt Institute for Botanical Documentation, Carnegie Mellon University, Pittsburgh, PA. Reproduced by permission of the Royal Tropical Institute, Amsterdam.

under W. F. R. Suringar, a taxonomist and opponent of evolution theory. However, he was strongly influenced by his contact with the young zoology professor Emil Selenka, an enthusiastic Darwinist, and by the works of Julius Sachs and Eduàrd Strasburger. On completing his doctorate, Treub stayed on to assist Suringar for seven years, during which time he published numerous papers on histology, cytology, and embryology. Then in 1880 he accepted the post of Director of the Botanic Garden ('sLands Plantentuin) in Buitenzorg (now Bogor) on the island of Java, which had been under Dutch control since the seventeenth century.[28] The 'sLands Plantentuin had been in existence since 1817, and its previous directors had built up the library and herbarium and had increased the physical size and collections of the garden. From the beginning, Treub was determined to make the Buitenzorg garden into a research facility for the study of tropical plants. In 1883–4 H. Graf zu Solms-Laubach, professor of botany at Göttingen, visited the botanical garden as Treub's guest. On his return to Germany, Solms-Laubach published an enthusiastic account of his experiences, in which he urged his fellow botanists to follow his example.[29] Using Solms-Laubach's visit as the catalyst, Treub succeeded in persuading the Dutch government to establish a botanical research station at Buitenzorg.

In preparation for the station, he converted a hospital into a laboratory, furnished it as best he could with modern scientific equipment, and revived

The combination laboratory and guest house of Melchior Treub's mountain research station at Tjibodas, in the highlands of Java, the site of frequent visits by German botanists in the 1890s. Reproduced by permission of the Royal Tropical Institute, Amsterdam.

the garden's journal, *Annales du jardin botanique de Buitenzorg*. In his report for the year 1884, Treub wrote:

> The visits which our Botanical Station will received from foreign scientists will not only serve to maintain the reputation of the 'sLands Plantentuin but, if my expectations are fulfilled, they will also provide our *Annales* with many important contributions to our knowledge of the life phenomena of tropical plants written by experienced authors.[30]

The station opened in January 1885, and Treub was not disappointed. From that time until 1909, when illness forced him to retire, 134 scientists visited the botanical station, commonly referred to simply as "Treub's laboratory." More than ninety botanists were included in that number, and they kept Treub supplied with abundant material for his journal.[31] Treub continually improved and expanded the research facilities at Buitenzorg, providing a laboratory environment that rivaled that of the botanical institutes at the German universities. He also added a mountain research station. When the 'sLands Plantentuin acquired part of a pristine rain forest at Tjibodas, in the highlands of Java,

Treub asked for permission to convert the former governor's holiday residence there into a small laboratory and guest house so that visiting botanists could carry on extended investigations in the montane forest. The Dutch authorities approved of his plan, and in 1891 Treub officially opened the mountain station of Tjibodas, which became one of the most popular attractions for foreign botanists.[32]

Germans were among Treub's most frequent and enthusiastic visitors. Thirty-eight German botanists in all made the trip to Buitenzorg between 1883 and 1909, over 40 percent of all botanists who visited the facility during that period.[33] This is not a surprising statistic. Treub and most of his staff spoke German, a rather important factor in attracting German visitors. The botanical programs at German universities also provided excellent preparation for the laboratory facilities at Buitenzorg. And, of course, there was the additional incentive of the proximity to recently acquired German protectorates in the Pacific. Few German botanists who were able to procure the means to travel could resist the temptation. Their institutions usually granted leaves of absence, and various scientific societies provided the necessary financial backing. With the exception of Heinrich Schenck (see Chapter 7), all of the young German botanists mentioned in this study visited Buitenzorg at one time or another. Haberlandt's opportunity came in the early 1890s.

HABERLANDT IN THE TROPICS

In the winter of 1891–2, when Volkens was planning his trip to East Africa, Gottlieb Haberlandt made his own botanical excursion to the Indo-Malaysian tropics. He was then professor of botany at the University of Graz, and his trip was sponsored by the Vienna Academy of Sciences. Since leaving Vienna in 1881, Haberlandt had spent several years teaching botany at a technical college in Graz and assisting botanist Hubert Leitgeb at the university there. In 1888, following Leitgeb's death, Haberlandt assumed the university chair. Although somewhat isolated in southern Austria, much as Schwendener had been in Basel, Haberlandt maintained regular contact with Schwendener and his students in Berlin, and he remained the most articulate spokesman for physiological plant anatomy. Nearly all of the more than thirty articles that he had published since his year with Schwendener dealt with some aspect of the relationship between plant structure and function.

The experiences in Java and Malaysia had a strong impact on Haberlandt. Like many of his professional colleagues, he was fascinated by his first opportunity to observe tropical vegetation in its natural setting. He carried on several research projects at Treub's laboratory and published extensively on his return to Europe: His most significant contribution to the study of plant adaptation in the tropics came in the form of a series of articles entitled "Anatomical-Physiological Investigations of the Tropical Foliage Leaf." The foliage leaf was the central focus of the biological studies represented by Schwendener's school; it is the locus of the biological function unique to plants, photosynthesis, and it serves as a convenient point of departure for discussing means of protection against the extremes of temperature and moisture. Haberlandt was struck by the wide spectrum of structural adaptations in tropical leaves, representing the effects of exposure to environmental conditions that, except for the extremes of cold and dryness, vary in the tropics over a considerably broader range than in the temperate zone. His series of articles covered everything from the shape and orientation of the leaves to the internal arrangement and structure of the photosynthetic tissue to the role of the hairs on the leaf surface.[34]

In a less formal contribution to the study of tropical vegetation, an amply illustrated anecdotal account of his journeys entitled *Eine botanische Tropenreise,* Haberlandt set down his views regarding the value of studying tropical plants in their native lands. He wrote from the perspective of pure science; unlike Volkens, he devoted little attention to the economic importance of tropical plants. He began by contrasting the old and new approaches to botanical excursions:

> The time has passed when a botanical excursion to the tropics was synonymous with a floristic-systematic journey of discovery, wherein the greatest reward for enduring toil and hardship was in finding as many new plant species as possible. The majority of scientifically trained botanists who visit the tropics today strive for an entirely different goal.

To begin with, Haberlandt continued, collecting itself has become considerably less rewarding:

> Whereas at the beginning of the century A. von Humboldt and Bonpland brought home no less than 3600 new species from approximately 5800 species collected during their several-years journey to the American tropics, at the end of the century one could be lost for days in the no less luxuriant and species-rich forests of Ceylon and Java without discovering but a few new phanerogam species.[35]

Yet over and above such practical concerns, the botanists of today have quite different preoccupations. Fifty years earlier, Matthias Schleiden had stated, in characteristic fashion, that the thorough biological investigation of a single group of Brazilian ferns would be as valuable as the discovery of 2,000 new species. Today, Haberlandt assured us, such a rebuke is not necessary:

> Along with anatomical and embryological investigations, the modern botanist also considers physiological experiments to be of value, and biological [here one could read "ecological"] research, which has made such amazing progress as a result of the stimulus provided by Darwin, finds an exceedingly fruitful field of observations in the tropics. It is, in a word, *general botany* which in more recent times governs the study of the tropical plant world.[36]

By "general botany" Haberlandt meant the study of plants as living organisms in their natural environments, as opposed to the examination, out of context, of isolated details of either structure or function.

Haberlandt added that studying plant life in the tropics serves also to correct misinterpretations or false impressions gained from restricting one's experiences to the temperate zone. Once one has experienced the year-round luxuriance of vegetation in the tropics, for instance, one realizes that the winter rest of temperate plants has come about as the result of bitter struggle against the adversity of climate. Further elaborating this theme, Haberlandt suggested that the tropical plant, rather than the temperate-zone plant, become the standard from which we derive our interpretations concerning adaptations to other zones, since plants have existed in the tropics for a longer period of time under continuously favorable conditions and have developed in response to a wider variety of external influences than have plants in other regions. Haberlandt was not the only European botanist to have this insight, of course; his views reflected a growing recognition among his temperate-zone colleagues that just as the study of tropical plants had shed considerable light on taxonomic questions, it would do the same for the problems of general botany.

Physiological plant anatomy did not bring about an immediate revision of botanical research methods. At Berlin in the 1880s it became a rallying point around which botanical students interested in studying the adaptive role of anatomical structures could focus their attention. This point of view was not without its problems. Centering one's attention on the function of every structural detail can lead to serious conceptual difficulties. Many botanists continued to resist the introduction of what they considered teleological considerations into their work, believing that Schwendener and his students placed too much emphasis on the concept of utility. Karl Goebel, for example, a contemporary of Haberlandt who gave considerable attention to the problem of adaptation in his own work, took a dim view of the practice of searching for the physiological significance of every anatomical feature. In 1886, having recently visited archaeological exhibits in Athens and Cairo, Goebel wrote to Julius Sachs: "Pity that Schwendener in his time had not been on the 'Creation Commission'; he would certainly have made everything more purposeful. As it turned out, plants received a huge share of purposefulness, and precious little was left over for man."[37] Although Goebel and others criticized the overuse of teleological interpretations, by the late 1880s the opposition, in principle, to physiological plant anatomy had died off considerably, and the general biological viewpoint that it promoted had surfaced elsewhere and was finding acceptance outside Berlin.

6

Beyond Schwendener's circle: Ernst Stahl

Laboratory-trained botanists were not in short supply in Germany in the late nineteenth century, and many of them had opportunities to supplement their training with travel to exotic and unfamiliar environments. Some of these botanists understood, and attempted to exploit in their own research, the relationship between Darwinism and the anatomical and physiological studies that, until the most recent past, had been restricted to the laboratory. Ernst Stahl and A. F. W. Schimper stand out as two of the most significant contributers to this research field outside the Schwendener circle; and to their names must be added that of Heinrich Schenck, Schimper's traveling companion and colleague for many years. All three were from the western borders of the expanding German state. Stahl and Schimper, both natives of Alsace, might never have studied or worked in German universities had it not been for the terms of the French surrender following the Franco-Prussian War. Schenck came from the vicinity of Bonn and remained there for much of his professional life. He and Schimper worked together under Eduard Strasburger at the University of Bonn throughout the 1880s and 1890s; Stahl spent most of his career at the University of Jena, where he went as Strasburger's replacement in 1881. Although Schenck offers a tenuous link to the Schwendener group, since he spent a year at Berlin in the early 1880s, he was influenced principally by Strasburger and Schimper. Stahl and Schimper, on the other hand, both studied for several years under de Bary, the old-school morphologist and plant pathologist who objected so strongly to the program of research carried out by Schwendener's students.

Anton de Bary (1831–88) had been one of the principal researchers responsible for the development of scientific botany in Germany during the 1860s and 1870s. He not only set up the first two botanical laboratories, at Freiburg and Halle, but his investigations into the life cycles of the fungi had led to fundamental insights into the fungus diseases of higher plants. Although, for Schwendener, Haberlandt, and many of the botanists at Schwendener's institute, de Bary may have epitomized the approach to the study of plants that they were trying to replace with physiological plant anatomy, he nevertheless continued to attract a large following. During the 1870s he established one of the most respected botanical research centers in the world at the reconstituted German university at Strasbourg, where students from England and the United States, as well as Germany, came seeking a thorough grounding in the fundamentals of plant science. These students were apparently undeterred by de Bary's strong emphasis on anatomical detail. Georg Klebs, Karl Goebel, George Karsten, Francis Darwin, and Gregor Krause, all of whom had interests that went well beyond the limits of descriptive plant anatomy, studied with de Bary at one time or another.[1] De Bary's students were no less exposed than those of Schwendener to the general intellectual trends of the 1870s and 1880s, and it is not surprising that some of them went on to direct their attention to problems of plant adaptation.

LEARNING FROM THE MASTERS

Ernst Stahl (1848–1919),[2] the oldest of the botanists considered in this study, was fortunate to have completed his training at a time when permanent positions were still to be found for young Ph.D.s. Good academic credentials, along with apprenticeship under both de Bary and Sachs, brought him a professorship at Jena relatively early in his career. The son of a lumber merchant from the Alsatian village of Schiltigheim, Stahl was raised and educated in nearby Strasbourg. As a result of his early interest in science, he became acquainted during his *Gymnasium* days with Wilhelm Philipp Schimper, the director of the Strasbourg Natural History Museum and father of A. F. W. Schimper. Stahl regularly accompanied Schimper and his son on botanical excursions, and it was from the elder Schimper that he first learned of Darwin's evolution theory. Stahl began his university education in 1868 at the Académie de Strasbourg, where he studied botany under Alexis Millardet. When war broke out two years later, he left for Halle, on Millardet's recommendation, to work with de Bary. Following the war, he returned with de Bary to Strasbourg. As a result of the annexation of Alsace-Lorraine, the German government, anxious to set an example for the French, reorganized the Académie de Strasbourg into the Kaiser Wilhelm Universität zu Strassburg and appointed a number of prominent German scholars to the faculty. The bulk of the appropriations for the new university went to the medical program, but all of the departments benefited from this nationalistic urge to outshine the French. De Bary, having been instrumental in setting up botanical institutes at Freiburg and Halle, was given the opportunity to do the same at Strasbourg in 1872.[3]

Although later Stahl is said to have adopted as his motto "My laboratory is Nature,"[4] during the 1870s his work kept him well within the confines of the

Anton de Bary. Reproduced by permission of Borntraeger, Stuttgart.

conventional laboratory. He received his doctorate at Strasbourg in 1873 for a paper on the anatomy and developmental history of lenticels – lateral openings in stems that function much like the stomata in leaves.[5] After completing this work, he remained in Strasbourg for another four years, pursuing, among other studies, a particularly fruitful investigation into the true nature of lichens. Simon Schwendener had recognized the dual (algal/fungal) character of these organisms some ten years earlier, but Schwendener's theory still required conclusive demonstration. A number of botanists worked on the problem during the 1860s and 1870s, and Stahl provided what was perhaps the most substantial verification of the Schwendener thesis. In his two-volume *Beiträge Zur Entwicklungsgeschichte der Flechten,* published in 1877, Stahl summarized the results of research carried out in Strasbourg in which (1) he had discovered the sexual reproductive structures of the fungus associated with a particular lichen, thus demonstrating conclusively its fungal character, and (2) he had artificially induced the growth of a lichen, complete with the characteristic fruiting bodies, by combining in the laboratory the fungal spores and the algal cells associated with it.[6]

The success of his lichen studies gave Stahl the confidence to advance to the next stage in his academic career: In 1877 he habilitated at Würzburg and became a *Privatdozent* under Julius Sachs.[7] Sachs quickly turned Stahl's attention from anatomy and developmental history to physiology, particularly the study of phototropism. The subject of plant tropisms in general was one of

Sachs's major research interests; he directed the work of a number of his students toward the investigation of plant responses to gravity, touch, chemicals, and especially light. Wilhelm Pfeffer, Hugo de Vries, and Francis Darwin, as well as Stahl, all studied various aspects of phototropism under Sachs during the 1870s.[8] While Stahl was a *Privatdozent* in Würzburg, Charles Darwin sent his son Francis there in preparation for assisting him with the experimental work for *The Power of Movement in Plants*.[9] Darwin's book cited Stahl's first publication on phototropism, a brief article on the effects of light on the movement of zoospores (motile, swimming spores in the nonvascular plants). A year later Stahl extended this work to include the desmids, a group of single-celled algae; and he followed this with a comprehensive study, summarized in a long article for *Botanische Zeitung*, on the effects of the direction and intensity of light on plant movements in general, including the orientation of chloroplasts in the higher plants.[10] A few months later, he published a further study in which he suggested that the pallisade parenchyma in leaves is adapted to light of high intensity, whereas the spongy parenchyma, usually situated just beneath the pallisade layer, is adapted to light of low intensity. He reasoned that in nature we should expect to find a differential development of the two tissue layers, depending upon whether a given plant grows under sunny or shady conditions.[11]

This was pure speculation on Stahl's part, but it represented the first glimmerings of his later all-consuming interest in plant adaptation. He did not have the opportunity to pursue this work in Würzburg. By the time the article on light intensity and the differentiation of photosynthetic tissue appeared in print, he was already back in Strasbourg, where he had taken a position as an assistant professor under de Bary. However, his return there lasted for only a year. In 1881, when Eduard Strasburger moved from Jena to Bonn, Stahl was offered the botany professorship in Jena. Upon learning of his appointment at Jena, Stahl's good friend Karl Goebel wrote to de Bary: "And I am very happy for Stahl that he will be offered a chair, yet I believe that his position at Strasbourg is more agreeable than that which awaits him in Jena beside Hallier [then an assistant professor in botany] and Haeckel."[12] Goebel's concern for his gentler friend was unwarranted. Stahl seems to have been quite happy at Jena and to have got along well with the faculty. Jena was a small university situated far from any urban center. The total student enrollment there was barely over 500 when Stahl arrived in 1881, and it was still under 700 by the turn of the century.[13] He enjoyed his new position from the start and grew to love the surrounding countryside, where he conducted much of his research. Stahl was a slight, frail, modest man who never married, had little interest in politics, and spent his leisure time reading philosophy and literature and studying music. Jena offered the perfect setting for a person of his disposition. He remained there for the rest of his life, and seems to have been popular with the students and even to have drawn an Alsatian following among them.[14]

Ernst Stahl. Courtesy of the Hunt Institute for Botanical Documentation, Carnegie Mellon University, Pittsburgh, PA.

When Carl Nägeli died in 1891, Stahl declined the offer of Nägeli's chair at Munich, and it went to Goebel instead. Unlike Schwendener, Stahl did not find the larger university, with its greater number of students and better facilities, a sufficient attraction to lure him from the quiet rural setting he had grown to love.[15]

At Jena, however, Stahl had little reason to feel the sense of isolation that had so bothered Schwendener in Basel. The reputation of the university more than made up for its small size. Jena was not only the university of Goethe (by association) and of Fichte, Schelling, Hegel, and Schiller, it was also an active center for biological science. In zoology, Carl Gegenbaur had established the science of comparative anatomy in Germany during his years at Jena (1856–73). Ernst Haeckel, very much influenced by Gegenbaur, had been on the faculty since 1862. Oscar and Richard Hertwig studied there with Gegenbaur and Haeckel, and Oscar remained in Jena as a professor of anatomy. In botany, Stahl had been preceded by Eduard Strasburger, Nathaniel Pringsheim, and Matthias Schleiden. If Darwinism had an academic center in Germany, it was certainly Jena. After 1860 Darwin's evolution theory served as the basis for Gegenbaur's comparative anatomy; Haeckel extended this work in his own research and became, as well, the foremost popularizer of Darwin's theory in Germany; Eduard Strasburger, influenced by Haeckel, became an enthusiastic proponent of Darwinism and advocated the application

tion of the phylogenetic method to all aspects of biology; and during Stahl's first decade in Jena, Arnold Lang became, through Haeckel's influence, the first Ritter Professor of Phylogeny.[16]

The Darwinian following at Jena emphasized the theory of descent, not natural selection. Gegenbaur virtually ignored natural selection in his work; Haeckel gave considerable attention to the concept in his general and speculative writing but did not pursue its implications in his own research; the Hertwigs focused their attention on cytology and embryology; and Lang concentrated on invertebrate phylogeny. Haeckel acknowledged that Darwin's natural selection theory necessitated the establishment of a new science – which he named *Oecologie* – to investigate the relationship of organisms to their "conditions of existence"[17]; however, his own work, and that of the other zoologists at Jena, stressed the use of the methods of comparative anatomy to investigate phylogenetic relationships (the term *Phylogenie* was another product of Haeckel's invention). Unlike his zoological colleagues, Stahl arrived at Jena after spending a number of years engaged in physiological research. During his year at Strasbourg, he had not returned to anatomical studies but continued the physiological work that he had begun with Sachs. Therefore, the Darwinian environment in which he found himself at Jena stimulated him not to pursue phylogenetic questions but to explore further the relationship between the plant and its habitat. In a *Festschrift* in honor of Stahl's seventieth birthday, Wilhelm Detmer, his longtime colleague and eventual successor at Jena, characterized Stahl's approach to his science as follows:

> The more Stahl became engrossed in the investigation of the living phenomena of plants, the greater became his interest in ecological phenomena, which he always considered from the standpoint of selection theory. For Stahl the individual plant is a unified system, exhibiting manifold correlative relationships among its parts, and influenced in the most diverse manner by external factors.[18]

This brief description might have characterized the scientific perspective of Haberlandt or Volkens as well. Like Schwendener's students, Stahl viewed the plant as an integrated, living system whose every detail reflects a close relationship to the external environment, the inevitable consequence of natural selection at work over countless generations. He developed this viewpoint contemporaneously with the Schwendener group and frequently cited their works, despite having been prepared by de Bary to take a skeptical view of physiological anatomy.[19] His move to Jena followed Schwendener's move to Berlin by only two years; and although he did not have the advantage of working with a team of colleagues under the banner of physiological plant anatomy, he certainly found at Jena a positive environment in which to pursue a very similar line of research.

It was at Jena that Stahl's research turned almost exclusively to studies of plant adaptation. Not a prolific writer, Stahl published only thirty-eight separate works in his lifetime. The longest of these barely exceeded 150 pages, and many were little more than brief notes. This was meager production by the German academic standards of his day; but, taken collectively, Stahl's works, especially those he produced at Jena, represent a fairly broad cross section of topics within the domain of plant adaptation. His papers covered subjects ranging from the meaning of dormancy in plants to means of protection against animals to the biological significance of calcium oxalate in the cell sap, but two general themes recur fairly frequently: (1) the effects of the light conditions of the habitat on the details of plant structure and (2) means of protection against external factors, from snails to excessive rainfall. There was a Darwinian undercurrent running through this pattern. Stahl, like Haberlandt and Volkens, was convinced that natural selection dictated that all plant structures serve an adaptive function. Since light is the most important factor in the habitat of the plant, Stahl reasoned that we should expect to find adaptations to both the quantity and quality of light reflected in almost every detail of plant structure. He was also convinced that one of the central concerns of Darwinian biology ought to be the study of means of protection against the excesses of the external environment. Stahl's Darwinism became increasingly explicit as his career at Jena progressed.

The first piece of research that Stahl undertook at his new position was a follow-up study to his brief speculative paper on the quality of light and the differentiation of photosynthetic tissue. He decided to make an extensive study of species throughout the plant kingdom, from liverworts and ferns to the flowering plants, correlating the light conditions of the habitat with the development of photosynthetic tissue. In the earlier article, Stahl had reasoned that since in horizontally oriented leaves the pallisade parenchyma lies near the upper surface and the spongy parenchyma near the lower surface, the former is adapted to light of high intensity and the latter to light of low intensity. Therefore, plants found normally in sunny habitats should display a well-developed spongy parenchyma. This pattern is exactly what Stahl's investigation turned up. Not only was he able to correlate the differentiation of photosynthetic tissue with the light conditions of the habitat, he was also able to show that under conditions of extreme shade, the pallisade layer is so underdeveloped in some species as to be indistinguishable from the spongy tissue.[20] He then went a step further and examined such structural features as the thickness of leaves, the nature of the intercellular spaces, the shape of chloroplasts, and the thickness of the epidermis. His findings contradicted an earlier statement by Haberlandt that the quality of light could affect the overall organization of the photosynthetic system but not the particulars of anatomical structure.[21] In contrast to Haberlandt's view, Stahl's comparison of similar species found in shady and in sunny habitats revealed that the intensity of light could affect such anatomical details as the shape and orientation of chloroplasts and the thickness of epidermal cells.

Except for brief remarks concerning the developmental history of photosynthetic tissue in selected species, Stahl declined to comment on the origins of the observed patterns of tissue differentiation. If we are to trust Wilhelm Detmer's retrospective analysis, Stahl believed that the degree of plasticity displayed by plants with respect to the anatomical details of photosynthetic tissue is a product of natural selection: The capacity of certain species to develop photosynthetic tissue in response to the light conditions of the habitat proved to be advantageous in the struggle for existence and became fixed over time in the hereditary makeup of the plants.[22] Although this view is certainly consistent with Stahl's later writing, his 1883 article made no explicit reference to evolutionary origins. However, his Darwinian position became clear in an article published five years later on protective plant devices against snails.

PLANTS AND SNAILS

"Pflanzen und Schnecken," published as an article and also as a separate work in 1888, carried the subtitle "A Biological Study Concerning Means of Protection in Plants Against Damage by Snails."[23] In his introduction to the snail study, Stahl expressed surprise that although Darwin's evolution theory had inspired research into the role of animals in the dissemination of pollen, fruits, and seeds, almost no work had been done concerning the destructive effects of animals on plants and the means of protection against them:

> Yet it follows from the outset that just as in the one case plants have developed the most diverse lures, in the latter case they must have developed protective devices against attack by animals, and that, in particular, every plant must be provided with such means of protection through which, if it cannot resist the attack of the surrounding animal world, it can at least resist extermination.[24]

Stahl attributed the neglect of this problem to the common association of spines, thorns, toxins, and other clearly defensive devices in plants with areas, such as steppes or deserts, where there is sparse vegetation and a relative abundance of herbivores. In other words, the need for protective devices against animals appears, on the surface, to be a localized phenomenon restricted to one or two climatic zones. However, Stahl argued, if striking defense mechanisms are lacking, for the most part, in native European plants, that does not mean that these plants are without means of protection against herbivory. Quite the contrary is true: "In general, a thorough investigation soon reveals that all the wild plants studied, even the seemingly defenseless, possess means of protection against attack by animals."[25]

Stahl consequently volunteered to make a modest contribution to help fill this gap in our knowledge of animal–plant relationships. On the basis of laboratory studies and extensive observations of the feeding behavior of snails in nature, he produced an exhaustive account of the protective devices in plants against these omnipresent herbivores. The means of protection that he described were either chemical, such as the production of sour sap, or mechanical, such as thick skin or bristly hairs. Every plant that Stahl examined possessed some means of protection against snails. Although he limited his study to herbaceous plants, he did not doubt that means of protection against snails are ubiquitous. Against the hypothetical objection that trees and shrubs do not require such protection, Stahl countered with evidence from Darwin:

> If this is true, with certain limitations, for full-grown plants, nevertheless we must not forget that in the seedling stage these plants experience the same dangers as the more lowly forms. The great harm which snails inflict upon trees and shrubs is familiar enough. Among 357 seedlings which had sprouted together in a small area, Darwin (*Origin of Species*, Chapter 3) observed that no less than 295 were essentially destroyed by snails and insects.[26]

Through careful observation in nature and through carefully controlled experiments in the laboratory, Stahl argued, we can determine whether or not a given plant is protected against a particular species of animal; and he outlined a procedure that involved a combination of field observations and laboratory experiments.

He admitted that observations in the wild are difficult to make and often yield quite variable results. In some cases, we can rely on the practical experience and the testimony of farmers, who, in a sense, have been conducting "experiments" for years on the food value of various plants; but we must exercise caution here, since animals in confinement will often eat a plant that they usually ignore in the wild. In any event, if it is clear that a type of animal either ignores, or only reluctantly consumes, a certain plant or part of a plant, then the next step is to determine the cause. This is easy if the plant possesses thorns or stinging hairs; but in cases where protection is achieved by means of a substance or substances within the plant, the exact nature of the defense mechanism is more difficult to determine. There is much room for error here, Stahl warned. His procedure was to observe in the laboratory the selective eating habits of snails exposed to plants with suspected chemical defenses and, in problematic cases, to attempt controlled experiments using chemically pure substances.

Once we have established that an animal or group of animals consistently avoids a particular plant or part of a plant, and we have also determined the mechanical device or chemical substance that is responsible, then, Stahl concluded, we are justified in considering that structure or substance a means of protection against the animals in question. How do we explain the origin and development of such means of protection? Stahl suggested two alternatives:

> Either one accepts that the devices in question – thorns, spines, stinging hairs, or the accumulation of tannins, ethereal oils, etc. – are acquired by the plants entirely independent of the surrounding animal world, and one therefore considers the sparing of such plants as merely fortuitous accident, or one sees in the devices under consideration the products of selection by the surrounding animal world. According to the first hypothesis, merely to cite an extreme example, the South African flora would have its thorny, glandular, bitter-tasting plants even it its multitude of herbivorous animals did not exist or had never existed.[27]

Such a conclusion seemed preposterous, but, Stahl pointed out, Grisebach tried to argue just that position. Always interpreting plant structure in terms of climate, Grisebach wrote that thorns and spines associated with vegetation in regions such as the African steppe are adaptations to dryness, since they help reduce evaporation by diminishing the surface area of epidermal tissue.[28] Stahl argued that these mechanical structures may be adaptations to the climate, but only indirectly, as means of protection against the abundant animal populations in such regions. Today, Stahl asserted, the majority of scientists agree that such devices have come about in causal connection with the influence of the animal world, and they consider the spines, thorns, bristles, and stinging hairs as "selection products of the animal world in existence either at present or in an earlier period."[29]

The situation is no different for the chemical means of protection. Little was known about the chemical role within the plant of such substances as tannins, ethereal oils, and alkaloids. Stahl admitted that each of these substances may well be an important link in some metabolic process. He also assumed that these substances were present in the plants before they became objects of selection by animals, but he added:

> Their present quantitative development, their distribution in the plant organs, their frequent peripheral location, and especially their early appearance, can only be understood as the effects of the animal world surrounding the plants. Still more, we cannot reject from the start the idea that even the quality of the excretions, in reference to their smell, taste, toxicity and, consequently, their chemical composition, must be influenced by the surrounding animal world, since, indeed, variability in plants may be assumed to apply just as well to products of metabolism as to form.[30]

Just as, through cultivation, man has selected an immense variety of good-tasting fruits from among unpleasant or tasteless wild types, we may assume that the selective activity of animals is responsible for the wide assortment of unpleasant and harmful substances found in plants. Here, as in many of his works, Stahl made use of one of Haberlandt's insights, in this case a suggestion in *Physiologische Pflanzenanatomie* that in some plants glandular secretions represent means of protection against snails and insects.[31] The diversity of these chemical substances, as well as the diversity of mechanical protective devices, Stahl concluded, "no longer appears to us as meaningless, but is just as comprehensible as the structural diversity of flowers."[32]

In an otherwise glowing eulogy for his old friend, Karl Goebel, ever skeptical of teleological arguments, pointed out the circularity in Stahl's reasoning. Assuming that plants possess means of protection against snails because they need them to survive, Goebel argued, Stahl was quick to categorize thorns, bristles, tannins, sour sap, volatile oils, alkaloids, and a host of other structures and substances as means of protection, although he did not carry out specific experiments designed to determine the exact relationship between each of these structures or substances and the behavior of snails.[33] Goebel's criticism was perhaps a bit harsh. It was true that Stahl's investigation had not established any particular plant substance or structure as, beyond any shadow of a doubt, a protective device against snails; but he also made it quite clear that he was well aware of the conceptual problems involved with identifying an organic substance or structure as a means of protection. For the most part, his investigation entailed observing, in the laboratory and in the field, the selective herbivory of snails toward plants with known physical structures or chemical substances that might be conceived as means of protection. In the article he carefully reported his observations and made strong implications as to the protective nature of the structures or substances in question. He took the position that, considering the devastation of which snails are capable, plant species lacking means of protection against snails would not have survived; and it follows that however these protective substances or structures may have originated, they have been preserved through natural selection.

STAHL IN JAVA: THE PURPOSE OF DRIP TIPS

Stahl conducted the field research for his study of plants and snails, as for most of his plant adaptation studies, in the vicinity of Jena. However, one year after publishing "Pflanzen und Schnecken," he had an opportunity to visit Treub's laboratory in Java. He spent over four months during the winter of 1889–90 at the Buitenzorg laboratory and botanical garden and at Treub's new mountain research station at Tjibodas. The main subject of his research was an investigation into the relationship between leaf shape and excessive rainfall in the tropics. Like Haberlandt, who visited the laboratory two years later, Stahl was struck by the diversity of adaptations in foliage leaves to the varied and often extreme conditions in the tropical environment; but whereas Haberlandt investigated the internal structure of the tropical foliage leaf, Stahl concentrated his attention on the external form of the leaf. In particular, he considered the significance of the exaggerated, drawn-out, pointed tips of tropical leaves (*Träufelspitze* – drip tips or drip points) as adaptations for facilitating the rapid drainage of water from the leaf surface.

As in the case of protective devices against snails, Stahl was once again faced with the task of demonstrating that a particular organic structure is, in fact, an adaptation to a particular aspect of the external environment. In his long report "Regenfall und Blattgestalt," published a few years later in Treub's journal, he approached this problem from several angles.[34] To begin with, simple observation had shown him that during a heavy rainfall a continuous stream of water is directed along the leaf surfaces and down the long tips. As a consequence, the upper surfaces of the leaves of plants equipped with drip tips were dry shortly after a heavy rain, whereas the leaves of native European plants growing in the Buitenzorg garden were still heavily laden with water. As an experiment, Stahl clipped off the leaf tips of selected tropical plants and discovered that, following a rainfall, leaves that were so altered remained wet for a much longer period of time than undisturbed leaves. He reasoned that the exaggerated tips help dry off the leaf surfaces not only by facilitating dripping but also by quickly transferring water droplets away from the broad portion of the leaf before they fall off. Further evidence to support his thesis came from the distribution patterns of plants with drip tips. For the most part, plant species equipped with long, drawn-out leaf tips are native to the humid tropics; and Stahl pointed out that the few indigenous European species with well-developed drip tips inhabit only the wettest environments in their temperate homes. As a third piece of evidence, Stahl noted that some tropical plants have leaves with a velvety upper surface (a particularly effective surface for repelling water) as well as the characteristic drip tips. Since such plants are found only in the dampest and shadiest habitats on the floor of the rain forest, Stahl concluded that this correlation between leaf surface texture and the presence of drip tips serves as indirect evidence for the water-draining function of the latter.

Having tentatively established the function of the drip tips, Stahl asked: What purpose is served by the rapid removal of rainwater from the leaf surface? What process is expedited by this action? From what dangers is the plant thereby protected? Stahl realized that such lines of questioning are often criticized as speculative and teleological, but he insisted on their value. Errors are likely to be made, he admitted, but Darwin's work conferred a new respect on inquiries into the purposes of organic structures; and although this area of research has yielded little so far in the way of positive results, that situation will likely change as biologists begin to make use of detailed comparative studies supplemented by experiments.[35] With regard to the particular example under consideration, Stahl offered a number of possible explanations for the adaptive significance of the draining function of the drip tips: (1) relieving the leaf surface of the heavy weight of the rainwater, (2) directing water to the roots of the plants (since leaves with drip tips tend to hang down), (3) cleansing the leaf surface of fungi, algae, lichens, and excrement from small animals, and (4) expediting the process of transpiration.[36] With his extensive knowledge of physiology, Stahl opted for the last alternative. Plants require continual transpiration from their leaf surfaces in order to draw up water and nutrients through their root systems. In the most humid tropical habitats, where many plants seldom receive direct sunlight and where there is danger that excess water on the leaf surface may hamper or prevent transpiration, Stahl reasoned that it is likely that plants would have evolved some means for keeping their leaves relatively dry.

Although he did not claim that this line of reasoning constituted proof that drip tips are adaptations for facilitating transpiration, Stahl believed that the exaggerated form of these structures in tropical vegetation helped give him insights into their function and that a comparison of other aspects of tropical plants to their counterparts in temperate regions might yield insights into the adaptive significance of many features of native European vegetation. He also maintained that in all such cases the adaptive structures in question are products of natural selection:

> Just as the often exaggerated form of the long drip tips of much tropical foliage provides us with the key to understanding their function in that location, the comparative observation of tropical plants, in which adaptations to rainfall are particularly evident, promises to grant us a deeper insight into the significance of rainfall for the development of leaf form in instances where it is less sharply imprinted, as with our native plants. It is far from our wish to attribute to this factor a *direct* role in the determination of form, in the manner of a *direct effect,* in Nägeli's sense. We consider the role of this factor, as well as others, restricted to the selection of adapted variations, the sifting activity of which has allowed only those forms to survive which have adapted to the rainfall in one way or another.[37]

Stahl was referring here not to Nägeli's own Lamarckian views but to his general criticism of such views. Nägeli was skeptical of the role played by the environment in any sense (Lamarckian or Darwinian) in the origin of adaptations, but Stahl clearly sided with the proponents of natural selection.

Stahl's trip to Java inspired two other works, one on the adaptive significance of colored foliage leaves and another on sleep movements in tropical plants (changes in leaf position at night).[38] Except for a brief botanical excursion to Mexico with George Karsten in 1894, Stahl confined his field research to the vicinity of Jena. In addition to plant adaptation studies, he occasionally published papers on purely physiological problems, but he did not hesitate to make speculations as to the broader significance of the phenomena under investigation. Particularly notable among such studies was his 1894 paper on transpiration and photosynthesis, in which he summarized much recent work (including that of Haberlandt, Schwendener, Tschirch, Volkens, and Schimper) on the interrelationship between these two important processes, and he also introduced a new method, the cobalt-paper technique, for detecting local variations in transpiration rate over the leaf surface. The technique consists of placing a piece of filter paper, which has been saturated with a solution of cobalt chloride and then dried, against the surface of the leaf. The dry blue paper turns pink on exposure to moisture and can detect minute differences in the rate of transpiration. With this technique, Stahl was able to demonstrate the far greater escape of water vapor from the underside of the leaf than from the upper surface and thus add another piece of evidence to support the view that stomata (which are usually more plentiful on the lower surface) play a significant role in both transpiration and photosynthesis.[39]

Many of the themes to which Stahl addressed himself are quite similar to those addressed by Schwendener's students during the 1880s and 1890s. Like Haberlandt, Tschirch, and Volkens, Stahl was motivated by his belief that natural selection is the central causal agent behind adaptive phenomena. In his lengthy assessment of Stahl's significance as a biologist, Wilhelm Detmer, his colleague at Jena since the 1870s, emphasized repeatedly that Stahl was a Darwinist in the strictest sense; he generally rejected Lamarckian interpretations of adaptation and he embraced most of Weismann's views even after 1890, when they came under sharp criticism in Germany.[40] The sources of Stahl's Darwinism are not easy to trace, since he wrote so little concerning the influences on his work. However, his experiences at Jena cannot be discounted, since neither the ecological nor the Darwinian orientation of his work surfaced in any definite manner until his Jena years.

With respect to Jena, his association with Haeckel cannot be taken lightly, although Stahl was more of a strict selectionist than was Haeckel. The two men could not have been more different in either temperament or scientific style; yet, despite Karl Goebel's cautionary remarks to de Bary on learning of Stahl's Jena appointment, Stahl and Haeckel became congenial colleagues, if not close personal friends. Haeckel's outspoken public manner, his polemical crusade for Darwinism, and the breadth of his speculative thought may well have seemed overbearing to the more cautious, soft-spoken Stahl; but perhaps he found these traits interesting and amusing, a welcome contrast to his own style. Both men shared the late-nineteenth-century faith in materialism and empirical science, as Otto Renner indicated in the following episode from his history of the Jena Botanical Institute. For many years, Stahl's apartment above the institute served as a frequent site of the "lecture evenings," organized by Haeckel, during which a number of Jena professors met for lively discussions. The serious discussions were usually followed by entertainment, which consisted of group singing aided by Stahl's piano accompaniment. The anatomists at Jena regularly attended these sessions, and Renner relates an unusual bequest: "During their lifetimes Haeckel and Stahl bequeathed their brains to the anatomists upon their respective deaths; and when the time came [both men died in 1919], Mauer faithfully preserved and carefully described these dissimilar receptacles of dissimilar souls."[41] It would not be unreasonable to assume that the most outspoken Darwinist in Germany, the man who coined the word "ecology" within a Darwinian context, exerted some influence on the evolutionary and ecological perspective of the colleague who joined him in willing his brain to science.

7

Schimper and Schenck: from Bonn to Brazil

Haeckel's influence may also have worked indirectly upon Stahl's younger Alsatian friend, A. F. W. Schimper, through Haeckel's former student and colleague Eduard Strasburger, under whom Schimper worked for many years at the University of Bonn. Schimper and his colleague Heinrich Schenck, coworkers under Strasburger, were also influenced strongly by the intrepid Darwinist Fritz Müller during an excursion to Brazil. Like Stahl, Schimper earned his doctorate under de Bary, but his interest in adaptation emerged only after he left Strasbourg. Yet whereas Stahl became a full professor at the age of thirty-three, Schimper was never in a position to carry out his research from the security of a professorship. Until close to the end of his life, he had to settle for docentships and assistantships that were not much to his liking. He traveled much more extensively than Stahl, partly because of his research interests, but also because of his unfulfilling and insecure academic positions. His career may have had an entirely different outcome, however, had the governing commission of a natural history museum not decided to reverse one of its decisions.

SCHIMPER'S CAREER: EARLY SUCCESS, DELAYED REWARDS

Andreas Franz Wilhelm Schimper (1856–1901)[1] inherited a family interest in natural history, particularly botany. His father, Wilhelm Philipp Schimper (1808–80), taught geology in Strasbourg and directed the natural history museum. Among his contributions to botany was an eight-volume study of European mosses, in which he was the principal collaborator. Two of his father's cousins were also botanists. Georg Wilhelm Schimper (1804–78) was a soldier and African traveler who made significant early contributions to the flora of Africa, particularly Abyssinia. Georg's brother, Karl Friedrich Schimper (1803–67), a friend of Alexander Braun and Louis Agassiz during their university days at Heidelberg and Munich, made contributions in many areas of natural science but is best known, as is Braun, for his work on the theory of phyllotaxy, the explanation for the serial addition of new leaves on a stem in spiral fashion.[2] Schimper's mother, Swiss born Adèle Besson, was a botanist herself and took considerable interest in her son's scientific education. It is not surprising that the youngest Schimper also decided upon a career in botany, although he at first combined botany with geology, his father's special field. After the 1870–1 war, the elder Schimper had accepted a position teaching geology and paleontology at the new Universität zu Strassburg, despite his French sympathies, in order to remain in Alsace. He also retained his position at the natural history museum. His son attended the university from 1874 to 1878, working with both Anton de Bary and mineralogist P. Groth. Stahl was working in de Bary's institute on his lichen study at that time. Schimper received his doctorate in 1878 for a study conducted under both Groth and de Bary concerning protein crystalloids in plants, and he stayed on to work in Groth's laboratory for another two years.

In 1880 Schimper returned to de Bary's institute, where Stahl was now an assistant professor, and resumed his work in botany. A year of research under de Bary resulted in the publication of a seminal paper on the growth of starch grains in plants.[3] This was the result of a straightforward, thorough piece of microscopic investigation that shed considerable light on the origin of starch grains and their relationship to photosynthesis. Earlier researchers had concentrated their attention on the appearance of starch grains in or near the chloroplasts in the mesophyll (the interior of leaves), and they had subsequently identified starch as the immediate product of photosynthesis. Schimper discovered that starch grains are manufactured in all regions of the plant, including stems and roots; they are produced either in the chloroplasts or in specialized bodies, found anywhere in the plant, which he called "starch builders." By carefully following the development of starch grains in numerous plant species, and by varying the light conditions under which some of these plants were growing, Schimper concluded that starch is a secondary product of photosynthesis, manufactured from an intermediary substance either in the chloroplasts themselves or in the starch builders. Schimper was a strongly independent, even secretive, worker. British botanist Frederick Bower, who worked by his side in de Bary's laboratory throughout that year, reported later that he never knew what Schimper was working on. Schimper never showed him his drawings and always consulted with de Bary in private. Bower discovered the topic of his research only when Schimper later sent him a copy of the published paper![4]

The starch grain research established Schimper's reputation within his discipline at the age of twenty-four. He may well have continued to work with de Bary on this and similar topics; but when the paper on starch grains appeared in print, Schimper had already left de Bary's institute due to a sequence of events that permanently altered his career. His father died in 1880, leaving vacant the position of director at the Strasbourg Natural History Museum. The Museum Commission, a holdover from the days of French rule, announced the younger Schimper, who had served briefly as his father's assistant, as its choice for the directorship. The science faculty at the university, de Bary included, rose up in opposition to the appointment of so young and inexperienced a man to this post. The Commission gave in to the pressure from the faculty and withdrew its appointment; and Schimper, embarrassed by the incident, left immediately for Lyon, where he found a position in a botanical garden. Not finding French science to his liking, he soon returned home to Strasbourg; but he felt that he could not resume his work at the university, although he did not resent de Bary for his role in this matter.[5] When an offer came along to join the new Johns Hopkins University as a fellow, he decided to accept it in order to distance himself from his immediate problems.

At Johns Hopkins, Schimper had limited teaching responsibilities and mainly continued his research on starch grains. His tenure there, however, was short-lived; he resigned after a year due to his aversion to the use of cats as experimental animals in H. N. Martin's laboratory.[6] Yet during his year in America he managed to do some traveling that proved to be very influential for his future research. He made brief visits to Florida and the West Indies, where he observed tropical vegetation for the first time. He also spent the summer of 1881 studying insectivorous plants at the zoological laboratory in Annisquam, Massachusetts. After his return to Germany in 1882, his research turned to studies of plant adaptation. He conducted most of this research from his position at the University of Bonn, where he worked from 1882 to 1898 as an assistant, a *Privatdozent,* and eventually an assistant professor. One of his primary responsibilities there, not by his choice, was to teach pharmacognosy (the identification of medicinal plants) to medical and pharmacological students. He finally obtained a professorship at Basel in 1898, but he did not live long enough to enjoy it. He died in 1901 from diabetes compounded by a case of malaria that he had contracted as a member of the deep sea expedition on board the *Valdivia* in 1898–9.

SCHIMPER AT BONN

The move to Bonn marked a decisive turn in Schimper's brief but eventful career. The professor of botany there was Eduard Strasburger, who had just come from Jena. Strasburger (1844–1912),[7] a Polish national, had studied with Ernst Haeckel and Nathaniel Pringsheim at Jena in the 1860s before he returned to Poland in hopes of securing a permanent academic position. However, the Russian regime in power at that time denied such appointments to Polish citizens. Strasburger consequently returned to Jena to take Pringsheim's place when the latter retired to Berlin in 1869. He remained at Jena for the next twelve years and, through Haeckel's influence, eventually became

Eduard Strasburger. Reproduced by permission of Borntraeger, Stuttgart.

a full professor there. Also through Haeckel's influence, Strasburger became an enthusiastic evolutionist. For a general lecture before the Philosophical Faculty in 1873 he chose as his topic "On the Significance of the Phylogenetic Method for the Investigation of Living Beings." Here he advocated the adoption of the phylogenetic point of view in all aspects of biological research – not just in the broad comparison of taxonomic groups but also in the investigation of minute structural features.[8] Strasburger's main research interest was in cytology, particularly the investigation of the cell nucleus. At Bonn he established one of the most important cytological research centers in Europe; and in the mid-1880s he was one of the chief proponents, along with Oscar Hertwig and August Weismann, of the concept of the primary role of the nucleus in the transmission of hereditary information.[9]

Schimper came to Bonn in 1882 fresh from his first experiences in the American tropics. Although he made additional contributions to the study of starch grains, work that fit in well with Strasburger's cytological research, his interests soon shifted to the investigation of plant adaptation. Perhaps Strasburger's evolutionary views helped influence this shift of interest, although Schimper's experiences in America had already begun to turn him away from pure laboratory botany. More important than his evolutionary views was Strasburger's relaxed attitude toward research. He allowed Schimper the freedom to pursue his new interest in adaptation in whatever direction it took him. According to Heinrich Schenck, only the friendly influence of

Strasburger and the liberal research climate that he provided at Bonn prevented Schimper from accepting a position at an American university in 1889.[10] Except for some minor contributions to pharmacognosy and a section on the seed plants in Strasburger's botanical textbook, Schimper's publications from the mid-1880s on were dominated entirely by the study of plant adaptation, particularly the application of a knowledge of plant physiology to the understanding of plant distribution patterns.

His first contribution to this field came in 1884 as a result of his second trip to the American tropics. His earlier visit to Florida and the West Indies aroused his interest in tropical epiphytic vegetation to such an extent that he planned to return soon in order to study this vegetation in depth. He had his chance within a year. In late 1882 he procured the assistance of the Alsace-Lorraine government to accompany Friedrich Johow, then a botanical assistant in Bonn, on an eight-month trip that took them to Barbados, Trinidad, Dominica, and Venezuela. Schimper's published report, "On the Structure and Habits of West Indian Epiphytes," offered an extensive discussion of variation in the epiphytic mode of existence and its relationship to environmental conditions. He insisted that floristic plant geography alone cannot account for the extreme local variability in epiphytic vegetation, and he offered his own explanation for this phenomenon:

> The efficient causes of the geographical distribution of the epiphytic plants in the West Indian archipelago have been to some extent entirely independent of their biological characteristics and of their adaptation to particular living conditions. This is true especially as regards the dissimilarity of the epiphytic flora of various islands, which corresponds to analogous differences in the terrestrial flora. These differences, insofar as this is not a question of endemic species, are attributable to the unequal distances to the continent and variations in oceanic and atmospheric currents, as well as the routes of bird migrations. Within the limited area of a particular island, however, considering the means of dispersal available to epiphytes by widespread seed dissemination, it must have been exclusively, or almost exclusively, the dissimilarities in living conditions which, at all events, brought about the striking differences in the distribution of epiphytic vegetation.[11]

Schimper explained that the most important local environmental factors, as regards the diversification of epiphytes, are light and moisture – temperature playing only a minor role and the physical and chemical conditions of the soil no role at all. In essence, epiphytes have evolved to inhabit the light and moisture conditions at every level in a tropical forest – from the lower regions, where light is scarce but moisture abounds, to the topmost branches, where there is an abundance of light but constant danger of desiccation.

Schimper was concerned not only with the nature but also with the origin of adaptations. Although in the 1884 article he did not state directly that the various forms of epiphytic plants are products of natural selection, he made

A. F. W. Schimper. Courtesy of the Hunt Institute for Botanical Documentation, Carnegie Mellon University, Pittsburgh, PA. Reproduced by permission of Borntraeger, Stuttgart.

the point quite explicitly in his book-length treatment of the subject, published after his third trip to the American tropics.[12] In the later version he devoted more space to his evolutionary explanation for the current distribution patterns of epiphytes, making clear his belief that adaptation to ever higher elevations in the forest has come about gradually, through the agency of natural selection. According to Schimper's view, the epiphytic mode of existence developed first at the lower elevations, where moisture is abundant; then, from countless generations of these lower-level epiphytes, survivors eventually emerged that could successfully inhabit the higher, drier levels of the tropical forest.

Epiphytes are able to grow in the forks of limbs and in crevices on rough stems, where small amounts of humus collect. Schimper reasoned that competition for such places must have been limited initially to a few species, since epiphytic existence requires very specific characteristics that are not at all common to plants growing in the humid tropics. Seeds must be small in order to settle and germinate in tiny fissures in the branches of trees, and they must have some means of vertical dispersal, either by wind or by arboreal animals. Schimper found that the seeds of the plant families with numerous epiphytic representatives, notably the orchids and bromeliads, fulfill all of these conditions, orchids producing tiny, wind-blown seeds and bromeliads producing small berries that are eaten by birds, with the seeds then deposited on branches in the droppings. In addition, since epiphytes must obtain all their moisture from atmospheric precipitation, the plants likely to exploit this rather specialized habitat would require relatively little water and would produce many lateral roots. This combination of characteristics is neither widespread nor particularly advantageous in terrestrial species in the tropics. "Hence the number of species that could emigrate to trees was relatively small, and victory over competitors was dependent on conditions other than those prevailing on the ground."[13] Thus epiphytes gradually evolved to survive under conditions quite different from those existing on the dark, moist forest floor:

> In those species which no longer grew on the ground and therefore could persist as epiphytes only, those characters were naturally selected that were specially suited for existence on trees; they have been adapted to this. Especially was every characteristic that enabled an epiphyte to advance upwards towards the light preserved and further developed. In the first place, in this relation protective means against the loss of water are in question, for every step on the way from the base to the summit of a tree brings with it not only more light but also greater dryness. Epiphytes growing at the base of trees in a rain-forest are hygrophylous, those that occur on the highest branches are xerophylous. The whole matter gives the impression of a gradual ascent from the deep shade into the sunlight, from the damp cool air of the interior of the forest to the dry heat of the top of the forest.[14]

Schimper then suggested that the most drought-resistant, sun-loving species were able to escape the rain forest and inhabit open country, hence their appearance now in areas with marked dry seasons. From a few species that were able to gain a foothold in the crevices in the lower limbs of trees, therefore, epiphytes have been modified, through natural selection, into a very diverse group of plants, representatives of which are adapted to environmental conditions at every level within the rain forest and even to conditions in the nearby savannas.

Perhaps Schimper's explicit references to natural selection in this work reflected the changing climate in Germany with respect to evolution theory. Discussing the influence of external factors on plants, Schimper stated: "It should be expressly noted that, with Weismann, I do not consider these external factors as the direct cause of hereditary characteristics, and therefore of adaptations; their role is restricted to the selection of the fittest variations at any given moment."[15] That Schimper took the trouble to point out his agreement with Weismann's views is an indication of the then current controversy in Germany regarding natural selection. In the four-year interim between Schimper's article and his book on the epiphytes, Weismann had published several articles on the continuity of the germ plasm and its implications for evolutionary theory, Rudolph Virchow and Weismann had had two confrontations at national meetings concerning this issue, and Theodor Eimer had published his Lamarckian-orthogenetic refutation of Weismann.[16] In the later version of his study of epiphytes, Schimper may well have thought it necessary to indicate where his sympathies lay on this matter.

SCHIMPER AND SCHENCK IN THE TROPICS

There was perhaps another reason for Schimper's overt references to Darwin's theory in 1888. He did the research for the second work in Brazil, where he and Heinrich Schenck, a companion from Strasburger's laboratory in Bonn, were guests of Fritz Müller in the small German colony of Blumenau, on Brazil's southeastern coast. Müller, the expatriate naturalist who had been living in Brazil since 1852, had been one of the earliest German proponents of Darwinism. His 1864 book *Für Darwin* discussed the applicability of the theory of natural selection to the life cycles of the crustaceans, and in subsequent work he further explored applications of Darwin's theory to embryology. His brother Hermann, who did not follow him to Brazil, spent years investigating the relationship between insects and flowering plants, eventually publishing a comprehensive study of pollination.[17] The two brothers maintained a regular correspondence with Darwin and Haeckel, and Fritz corresponded with Weismann as well. When Schimper and Schenck arrived in Blumenau, Müller immediately took the young botanists under his care and set them to work on two of his pet projects. Schimper worked with Müller on an investigation into the symbiotic relationship between ants and *Cecropia* trees; Schenck examined the adaptations of tropical lianas to their climbing mode of existence. Both investigations resulted in major publications.[18] If the visit to Blumenau had a positive influence on Schimper and Schenck, it also made quite an impact on Müller. His biographer, Alfred Möller, wrote: "For years Fritz Müller recalled with visible warmth the glorious months that he spent in zealous wanderings with the two young friends, who had become quite dear to him; and to the end of his life he followed their fortunes and their works with genuine interest."[19]

In planning their excursion to the tropics, Schimper and Schenck could not at first decide whether to go to Brazil or to Cameroon, which had become a German protectorate in 1884. In considering the greater dangers involved in a trip to the lesser-known African territory, Schimper wrote to Schenck from Alsace in 1885:

> The matter of climate is serious. In Grisebach [*Die Vegetation der Erde*], vol. II, p. 112, I found the following sentence: "How general the perniciousness of climate may be follows from the fact that by far most scientific travelers in the most diverse landscapes become carried away with the experience, but in the tropical lands, almost without exception, well-known scientists are happy to return home." To die for science is, to be sure, no less *dulce et decorum* than *pro patria mori*, but I wish not merely by my death, but also by my work, to earn a name for myself in the history of science.[20]

There is a sad irony in these remarks, considering that Schimper contracted the illness that eventually took his life while off the coast of Cameroon on the *Valdivia* expedition. For the moment, however, Grisebach's words of caution were sufficient to dissuade Schimper and Schenck from choosing an African itinerary. On the eve of his departure, Schimper wrote to Daniel Coit Gilman, president of Johns Hopkins University:

> I shall leave Bonn tomorrow for a short trip to Brazil. My leave of absence being a very short one, I shall not be able to spend more than three or four months in the great South American empire, the richest country in the world for a botanist. It will be my third trip to tropical America; I have been twice in the West Indies and in Venezuela and I was both times carried away by most frantic enthusiasms about the magnificence of the vegetation.[21]

Schimper and Schenck left for Brazil in August 1886, spending September through November with Müller in Blumenau. Schimper had to return to Europe in mid-December to resume his teaching duties, but Schenck stayed behind and visited a number of sites in Brazil before he returned in July 1887.

For Schenck, this was his first and only excursion to the humid tropics. The son of a doctor from Siegen, Heinrich Schenck (1860–1927)[22] first came to the University of Bonn as a student in 1880. In 1881–2 he studied briefly under both Schwendener and Eichler in Berlin, but he returned to Bonn and took his Ph.D. under Strasburger in 1884. His dissertation, following the cytological emphasis at Strasburger's institute, concerned the thickening of cell walls in plant hairs and epidermal tissue. Immediately after completing his doctorate, however, Schenck began an investigation of the special adaptations of water plants to their submerged existence. His two treatises on the subject follow closely the spirit of the early work of Haberlandt, Tschirch, and Volkens; and it is likely that during his brief period in Berlin, Schenck fell under the influence of the Schwendener school.[23] He remained at Bonn as an assistant and a *Privatdozent* until 1896, when he found a permanent position at the Polytechnical Institute in Darmstadt, where he spent the rest of his career. His duties there included managing the botanical garden, which expanded considerably in both size and scope during his tenure. Particularly notable were his additions of several species of cactus collected during an excursion to Mexico in 1908. Besides authoring his own works, Schenck became a valuable editorial assistant. When Schimper became ill in the late 1890s, Schenck not only helped edit his plant geography text, he also saw to the publication of the results of Schimper's investigations of island vegetation conducted on the *Valdivia* deep-sea expedition. In addition, he contributed several volumes to George Karsten's *Vegetationsbilder* and assisted in editing the collection. He also published the results of the botanical collections of the German Antarctic Expedition of 1901–3.[24]

Heinrich Schenck. Courtesy of the Hunt Institute for Botanical Documentation, Carnegie Mellon University, Pittsburgh, PA. Reproduced by permission of Borntraeger, Stuttgart.

Plants and ants

The first project that Schimper undertook in Brazil was the study of a symbiotic relationship between ants and trees.[25] The imbaúba tree, *Cecropia adenopis,* in southern Brazil is always inhabited by the same species of ant, *Azteca instabilis,* a particularly ferocious species that protects the tree from leaf-cutting ants. Schimper and Müller found that any *Cecropia* tree not inhabited by these ants had its leaves bitten down to their midribs by the leaf cutters. *Azteca instabilis* inhabits cavities in the stems of these trees, a feature common to many trees in that region:

> The cavity, and therefore the dwelling place of the ants, in spite of its great utility, is not an adaptation to the guests; it represents rather a feature that is common to many other plants, and may be explained by the mechanical principle of construction as being the method of producing the greatest resistance to bending with the least expenditure of material. The dwelling existed before the symbiosis. It is otherwise with the entrances to it. Here indubitable adaptation is exhibited.[26]

The entrances to which Schimper refers are external depressions in the internodes of the stems corresponding to similar depressions in the internal walls. These depressions come at the ends of shallow vertical grooves formed by the pressure of axillary buds during the longitudinal growth of the internodes. Unlike the grooves, which contain lignified tissue, the depressions lack any tough or hard elements and are so thin as to appear as mere diaphragms.

Although similar grooves are found in the stems of many related species, the diaphragmlike depressions are unique to the inhabited trees. Schimper offered this explanation:

> At the phylogenetic commencement of symbiosis, the ants bored an entrance through the groove, evidently because the wall was somewhat thinner there, and in particular, in accordance with a custom that is almost always followed and is connected with the domestic arrangements, they bored as much as possible in the upper part of the dwelling. All features that facilitate boring through this place must have been retained in the struggle for existence, and been further added to through selection. They finally led to the differentiation of the thin, weak diaphragm that has been described.[27]

In exchange for protection, the ants receive not only a dwelling place but also food, in the form of whitish bodies rich in protein and oils found at the base of the leaf petioles. "Müller's corpuscles," as these bodies are called, after their discoverer, appear to be metamorphosed glands. However, they never perform glandular functions, and unlike true leaf glands, they are produced continuously during the entire life of the leaf. They are shed when they become gorged with albuminoids, and the ants take them away. As Schimper explained, "*Cecropia* affords an exceptional case of the voluntary surrender by the plant of proteid substances even in relatively large quantity, for Müller's corpuscles are produced continuously and in profusion."[28]

As a final piece of evidence for the efficacy of a Darwinian explanation for these structures, Schimper offered the contrasting example of an uninhabited, but closely related, tree species:

> The assumption that the entrance-door and Müller's corpuscles represent adaptations to ants was surprisingly confirmed by the discovery in the Corcovada, near Rio de Janeiro, of a *species of Cecropia devoid not only of the ants but also of the entrance-door and of Müller's corpuscles.* [Schimper's emphasis][29]

This species has the shallow groove in the internodes and even a depression at the upper end of the groove, but the depression is not nearly as pronounced as in the inhabited species, and the lignified tissue in the groove continues unchanged into the depressed area. Furthermore, this tree is not injured by leaf-cutting ants and seems to derive its protection from a waxy coating on the stem that impedes climbing. Schimper concluded that the entrance door and food bodies found in the inhabited species thus represent splendid examples of sophisticated adaptations to animal guests that, when examined in detail, lend themselves quite nicely to a Darwinian explanation.

Schimper's was not the only explanation. In a work published in 1910, American entomologist William Morton Wheeler objected to Schimper's characterization of the ants as protectors of the trees. Wheeler, no champion of Darwinian selection theory, argued that leaf-cutting ants rarely visit the *Cecropia* trees and that the *Azteca* ants mainly attack alien colonies of their own species in order to protect their feeding grounds and their nests. He also denied that Müller's corpuscles represent an adaptation to the ants. They might serve some excretory function as yet undetermined and may only incidentally provide food to the ants. Finally, he pointed out that other species of *Cecropia* with very similar internal structures are not inhabited by ants. It seemed clear to Wheeler that although the ants make use of the hollow stems and Müller's corpuscles, their relationship with *Cecropia* is not truly symbiotic and no coevolution has taken place.[30] Schimper may well have selected his evidence carefully to support his Darwinian explanation. In fairness to Schimper, Wheeler cited more recent work of which Schimper could not have been aware, and he was predisposed not to accept an explanation based on natural selection.

The anomalous stems of lianas

While Schimper was working on his studies of ants and epiphytes in Brazil, Schenck was gathering information for his study of the biology and anatomy of lianas. The resulting 500-page work, which took him six years to complete, presented a comprehensive discussion of the manifold adaptations and numerous representatives of this plant group, whose central unifying characteristic is dependence for vertical support on the stems and branches of other self-supporting plants. In the first part of this work, Schenck discussed the various modes of existence of lianas, breaking them into four groups – scramblers, root climbers, twiners, and tendril climbers – depending upon the method by which they attach themselves to their hosts. Here Schenck followed the pattern presented in Darwin's *The Movements and Habits of Climbing Plants* (1876). He then offered a complete taxonomic survey of the four groups as they are represented in the Brazilian flora. In the second part of the book, Schenck provided a detailed discussion of the peculiar anatomical feature of lianas: the presence of numerous separate woody cylinders in the stem, in contrast to the single central woody core in a normal stem. This arrangement, the so-called anomalous secondary thickening of lianas, provides strength to the stem and support to the long conducting tubes and vessels necessary for conveying nutrients over great distances.

By addition of concentric layers of lignified tissue, a normal stem develops a large central woody core suitable for providing nearly rigid vertical support. A woody stem that develops in the normal way cannot provide the necessary flexibility for a climbing mode of existence. The liana stem consists instead of woody strands interspersed with areas of soft tissue, the separate strands of wood acting, in effect, like the separate strands of a cable.[31] The rather unusual structure of the liana stem was well known at the time of Schenck's investigation. Fritz Müller had attempted to explain its physiological significance as early as 1866, but no one had yet attempted a full-scale investigation of lianas.[32] Schenck found the anomalous secondary stem thickening to be ubiquitous in lianas, and he found that the anomalous thickening does not follow simple patterns corresponding to either the taxonomic group or the mode of climbing, nor does it arise in the same manner in all liana species. In some cases, the cylinder of vascular tissue begins to develop in the normal fashion and later breaks up into sections; in others, several separate cylinders are present from the beginning; in still others, additional cylinders develop outside the primary cylinder, and there are many variations on these three basic themes. The result, however, is always the same: a stem with several separate strands of woody tissue that are able to slide past each other like the separate strands of a cable.[33]

To account for the helter-skelter distribution of these anatomical features, Schenck appealed to an evolutionary explanation. Since tendril climbers, scramblers, twiners, and root climbers are found in a wide range of taxonomic groups and in many regions of the world, he reasoned, it follows that the peculiar anatomical features of lianas have developed from a diversity of raw material and under varied environmental conditions. We should not, therefore, expect to find a simple correspondence between mode of existence and anatomical detail. The breakup of the central woody cylinder must have developed independently in many species from variations on the normal stem as an adaptation to the climbing mode of existence. This explanation was, of course, based upon natural selection:

> The complex types have evolved as useful modifications from simpler variations through the action of natural selection; and these simple variations, such as furrowed wood bodies [cylinders], may have appeared first without somehow being of use to the plant. Although the form of thickening may have proved to be quite variable at one time, natural selection was able to take hold and develop an anomaly further in a given direction.[34]

The extreme variation in the form of the anomalous thickening of liana stems is thus the result of natural selection acting upon the chance appearance, within different taxonomic groups, of plants with suitable stem forms. In order to show that these modifications of structure lie within the general range of variability of the plants in that region, Schenck took pains to describe nonclimbing Brazilian species that have furrowed central cylinders or other variations on the normal woody stem that might suggest a proto-liana stem modification. Under the continuously favorable environmental conditions in the tropical rain forest, lianas could have evolved from any group capable of generating individuals with woody tissue that might lend itself to the climbing mode of existence. Among the twining form of lianas alone, for example,

Schenck found representatives of no fewer than sixteen separate families, including hundreds of species, in the Brazilian flora.

The Indo-Malaysian strand flora

Schenck's study of lianas appeared as two volumes in a nine-volume series edited by Schimper under the general title *Botanische Mittheilungen aus den Tropen*. Schimper's study of ants and *Cecropia* trees and his work on Brazilian epiphytes appeared as part of this series, as did another of his works on the coastal vegetation of Malaysia. The last work was the result of Schimper's fourth excursion to the tropics in less than a decade. Within a few years after his return from Brazil, he secured the financial assistance from the Academy of Sciences of Berlin to spend several months in Ceylon and Java. The main destination for Schimper was the Buitenzorg Botanical Garden in Java and the laboratory of Melchior Treub. The excursion extended from the late summer of 1889 to the spring of 1890, and for part of this time Schimper was accompanied in Buitenzorg by Ernst Stahl and Alexander Tschirch. On this trip Schimper encountered a hazard for which even Grisebach had provided no warning. En route from Ceylon to Buitenzorg, he was detained for a considerable time in Batavia, because customs officials there had received word by telegraph of the arrival of a large shipment of opium. Taking Schimper for the expected opium smuggler, they subjected his luggage to a meticulous search before satisfying themselves that the contents were, in fact, harmless and legal.[35]

Once safe in Buitenzorg, Schimper, like all of the Europeans who visited the famous garden, was struck by the diversity of Javan vegetation and by the extremely advantageous location of Treub's laboratory. In future publications he heaped praise upon the laboratory for the role it played in advancing the study of plant adaptations. For his research he chose to investigate the adaptations of strand plants to their saltwater environments. He issued two published reports on this research: a short speculative paper, in which he presented the central problem of coastal vegetation as the need for protection against transpiration, and a longer volume for the series on tropical botany. In the first publication, Schimper discussed the physiological and anatomical aspects of the problem and offered some speculations regarding the distribution of alpine plants in coastal waters.[36] The particular difficulty faced by strand plants is that of dehydration due to the scarcity of salt-free water. The most common anatomical adaptation of coastal plants to the dangers of desiccation is a thick, impermeable cuticle forming the outer wall of the leaf epidermis. The heavy cuticle is a characteristic of xerophytic plants as well, particularly those found at higher elevations, where strong insolation makes excessive transpiration a constant problem. Such considerations, Schimper explained, account for the presence in the coastal regions of Java of many plants found also in the mountains. He added that such an explanation applies also to the distribution of alpine plants in coastal regions of Europe, and, analogously, to the presence of evergreen trees in drier regions of the temperate zone, since the needles of these trees are equipped with heavy cuticles. Furthermore, he suggested that deciduous trees shed their leaves not as a response to temperature but as a response to the difficulty of obtaining water from the cold soil.

Schimper ended this fertile piece of speculation with a methodological note concerning the value and applicability of plant adaptation studies. In order for such studies to be of use, he insisted, one need not present a complete explanation for all of the physiological processes in question:

> In the preceding, little attempt was made to explain the means of protection against transpiration physiologically; at present every attempt to do so would lead only to insupportable hypotheses. Yet biologically, that is, in terms of their importance for the life of the plants, these features are completely comprehensible to us; and this essay should serve only to furnish evidence that all plants which must struggle continuously, or periodically, against impeded water supply – be the cause of the latter the dryness of the atmosphere and soil, strong insolation or rarefaction of the air, or the saltiness or low temperature of the substrate – possess such means of protection.[37]

Schimper, like Haberlandt, believed that with regard to the problem of adaptation, there is value in making speculations before all of the facts are in, that we can understand the adaptive function of a particular anatomical structure without comprehending fully the details of all of the physiological processes in which it is involved. This was typical of the anti-inductive approach of these Darwinian botanists. An earlier generation had compiled heaps of isolated anatomical and physiological facts without shedding much light on the problem of adaptation. The time was ripe now for synthesis.

This viewpoint did not prevent Schimper from attempting a fuller treatment of the subject of strand vegetation. The following year, 1891, he published *Die indo-malayische Strandflora,* in which he applied the principles outlined in his earlier paper to the general problem of the distribution of plants in the coastal areas of the entire Indo-Malaysian region. Following an introductory section on the relationship of the salty substrate to the structure of strand plants, Schimper discussed the various vegetation zones and presented, in a manner similar to that of his work on epiphytes, a lengthy treatment of the distribution of strand plants. Here he discussed the means of seed and fruit dissemination, the role of birds and wind, and the effects of ocean currents on the distribution patterns of coastal vegetation. The final section, "On the Evolution of the Indo-Malaysian Strand Flora," contained very explicit statements regarding Schimper's Darwinian perspective. The main point of this section was that the strand vegetation represented a unique collection of plants peculiarly suited, as a result of the effects of natural selection, to the conditions of existence on the coasts. "On the strand," Schimper wrote, "the conditions of the struggle for existence are different than in the interior; consequently, many of the first colonizers owed their survival in the early days to fortuitous conditions that enabled them to hold their own, and only in the course of time, through directional [*einseitig*] selection, have they acquired special adaptations to their new mode of life."[38]

Schimper did not elaborate upon this explanation but stated it here more or less as a general assumption. He took a strict selectionist approach to evolution, holding to the view that natural selection has been the central causal agent in the differentiation of species. Botanists since Nägeli had argued that, although natural selection may play a role in weeding out the least fit individuals, species characteristics are, by and large, not based upon adaptation. Against this view, Schimper offered numerous examples from his study of strand plants in which a coastal species exhibits a special feature that is entirely lacking in its inland relatives, such as fruits well suited for ocean transport or seeds that can take hold only under the conditions that are present among the mangroves. He added that many features of plants not previously considered adaptive have been found, on closer inspection, to have adaptive significance. He was referring here to the mass of literature by physiological plant anatomists over the previous decade linking details of plant structure – such as the arrangement of photosynthetic tissue and the orientation of stomata and chloroplasts – to specific functions relating to the natural environment of the plant. In this section, Schimper seems to have been responding to the host of criticisms of natural selection that appeared in Germany, mainly in opposition to the views of Weismann, in the late 1880s and early 1890s. In his concluding remarks, he stated that although new criticisms of Darwin's theory continued to appear, despite the evidence that he and others had presented, "it does not seem necessary to me to point out the indefensibility of the new cases at hand."[39]

PHYSIOLOGICAL PLANT GEOGRAPHY

If Schimper's Darwinism is evident in *Die indo-malayische Strandflora,* so also is his turn toward plant geography. Following the publication of this work, he suspended additional research projects for several years in order to amass in a single volume the results of studies, published over the previous two decades, that centered on the approach to the investigation of plant distribution that he chose to call "physiological plant geography." The poor condition of his health following his Asian trip may have influenced his decision to write this textbook. In 1891, shortly after returning from Java, he declined the offer of the chair in botany at Marburg, as Goebel's successor, due to poor health.[40] Yet whether health or other factors led him into this new project, Schimper's efforts culminated in the publication in 1898 of the work for which he is best known, *Pflanzengeographie auf physiologischer Grund-*

lage.[41] A decade earlier, in his work on Brazilian epiphytes, he had already called attention to a new approach to plant geography that takes as its starting point the interrelationships between plants and their surroundings. At that time he referred to the new approach as "biological plant geography," and he cited as examples, in addition to his own works, Darwin's *Origin of Species,* Grisebach's *Vegetation der Erde,* Schenck's *Biologie der Wassergewächse,* and Volkens's *Flora der ägyptisch-arabischen Wüste.*[42] In the later textbook, perhaps to confer an even greater sense of scientific respectability on the enterprise, he referred to the new approach as "physiological."

Schimper was well aware of the important impetus for this new direction in plant geography given by the recent opening up of tropical lands to scientific investigation. In an 1896 report on the progress of plant geography, he cited the colonial movement and the establishment of the botanical laboratory at Buitenzorg as two important sources of stimulus. Concerning the latter he stated:

> More precise knowledge of plant physiology has led in most recent times to a scientifically-grounded physiological direction in plant geography, for which the botanical laboratory at Buitenzorg has become the most important center. To be sure, much of the work undertaken there deals with purely morphological problems, but almost all that rests upon physiology provides building blocks for the broader framework of physiological plant geography. . . .[43]

A similar passage appears in the preface to his plant geography textbook. In Schimper's view, the opportunity for laboratory-trained botanists to conduct long-term studies in the tropics had led to a far more sophisticated approach to problems of plant adaptation than that provided by earlier plant geographers, who generally restricted themselves to investigations of broad-scale patterns of vegetation cover.

Schimper was quite familiar with Grisebach's work and made frequent references to *Die Vegetation der Erde* in his earlier writing. However, his approach to plant geography differed from that of Grisebach in two important respects. First, whereas Grisebach presented a rather static picture of vegetation patterns throughout the world, Schimper, writing from an evolutionary and selectionist point of view, conveyed a sense of continual transformation. In the preface to his textbook he stated:

> Existing floras exhibit only one moment in the history of the earth's vegetation. A transformation which is sometimes rapid, sometimes slow, but always continuous, is wrought by the reciprocal action of the innate variability of plants and the variability of the external factors.[44]

He added that this transformation is effected primarily through the selection of chance-appearing characters that arise by unknown means. A second point of departure from Grisebach concerned the use of plant physiology. Although Grisebach's work emphasized the relationship between vegetation and climate, it contained only simple and generally incomplete references to physiological processes. Schimper's work, on the other hand, stressed above all the physiological viewpoint. The following passage is also from the preface:

> The connection between the forms of plants and the external conditions at different points on the earth's surface forms the subject-matter of ecological plant geography, which has only recently become a prominent subject of interest, although it found a place in earlier works, especially in Grisebach's *Vegetation der Erde,* where, however, it was regarded from obsolete points of view. The greater prominence of physiology in geographical botany dates from the time when physiologists, who formerly worked in European laboratories only, began to study the vegetation of countries in its native land.[45]

Following this passage, Schimper reiterated the view, often expressed by members of the Schwendener school, that the study of vegetation in extreme environmental conditions, such as those present in the tropics and in desert and arctic regions, provides the best means for determining the exact character of specific adaptations. Such determination, Schimper asserted, is the objective of the physiological point of view.

Schimper's departure from the older approach to plant geography was not confined to the preface of his textbook. Although there are few specific references to natural selection in the body of the book, Schimper often implied selection theory quite strongly with such statements as "The absence of terrestrial halophytes in the nonhalophytic terrestrial flora is due only to their incapacity to struggle successfully; the absence of aquatic halophytes in fresh water depends on their unfitness to exist there."[46] The physiological point of view is clear and explicit throughout. The book begins with a long section (170 pages in the German edition) treating individual factors that affect plant distribution – water, heat, light, air, soil, and animals. Schimper examines each factor in its physiological detail, citing abundant examples from recent experimental work. Following a brief discussion of the major plant formations, somewhat in the style of Grisebach (50 pages), Schimper then turns to an over-600-page discussion of the relationships between plants and environment in each of five major vegetation zones – tropical, temperate, arctic, montane, and aquatic. Here he includes such topics as "metabolism and interchanges of energy in mesothermic plants at different seasons," "histological peculiarities caused by continuous illumination," and "individual periodicity of the separate shoots of many tropical plants." The textbook as a whole represents a direct application of recent ecologically oriented work in plant physiology to the broad problems of plant distribution. Schimper did not undertake any new research for this study. Rather, he made abundant use of his earlier work, as well as that of Schenck, Stahl, Volkens, Tschirch, Haberlandt, Schwendener, and a number of other researchers. The textbook represents, in effect, an encyclopedic overview of research concerning plant adaptation conducted during the last two decades of the nineteenth century.

Schimper's plant geography textbook is best known as one of the early formal contributions to the nascent science of plant ecology. In a sense, however, Schimper's 1898 work introduced no new methodological principles or points of view; rather, it drew attention to a body of research that had been in existence for many years. The botanists and plant geographers whose interests concerned more holistic themes provided the discipline-building impetus in ecology. However, Schimper and the other German botanists who brought physiology outdoors in the 1880s and 1890s were not interested in founding a new discipline; they were interested mainly in broadening the scope of scientific botany. Armed with the legacy of a quarter-century of somewhat narrowly conceived laboratory work in anatomy, morphology, and physiology, they chose to venture out into the natural habitat of the plant. They received encouragement for this venture from two sources: one political, the other intellectual. On the political side, a government that had recently entered the colonial arena generally encouraged overseas travel and provided botanists with opportunities to bring their research into environments that differed radically from any found in Europe. On the intellectual side, the relatively unexplored aspect of Darwinism, the investigation into the implications of natural selection for the problem of adaptation, cried out for attention and fit in well with the physiological orientation of their research. This second source of motivation deserves further attention, since it raised once again, and in a new form, the teleological questions that an earlier generation had abandoned.

8

Teleology revisited? natural selection and plant adaptation

It is one of the ironies of the history of the life sciences that the Darwinian evolution theory claimed to offer a complete explanation for the adaptedness of organisms to their environments, and yet the initial effect of Darwin's theory on the life sciences, in Germany as elsewhere, was to concentrate attention on the details of phylogenetic development rather than on the process of evolutionary change. Perhaps the explanation for this situation is that in much of the life sciences before 1859 adaptedness was taken for granted – as in natural theology or in the comparative anatomy of Cuvier – but descent from common ancestors was not. Establishing the fact of descent was at first the more significant contribution of Darwin's *Origin of Species;* natural selection was simply the plausible mechanism that helped elevate Darwin's version of evolution theory above those of Lamarck, Robert Chambers, or Geoffroy Saint-Hilaire. Darwin's allies as well as his enemies had reservations concerning the details of the natural selection process, its general applicability, and its efficacy as the primary agent of organic transformation; but it was at least a process that relied upon familiar phenomena and did not invoke strange new forces.

Yet the studies of plant adaptation that appeared in the 1880s found their source of inspiration not in the theory of descent but in the evolutionary mechanism proposed by Darwin. Darwinism, as natural selection theory, provided them with the justification for looking beyond the particulars of structure and function and examining the larger, and more elusive, question of purpose. Natural selection entailed a causal, mechanistic, materialistic explanation of the process of speciation; it conferred a new legitimacy on teleology, albeit a somewhat restricted interpretation of teleology. In contrast to evolutionary views that emphasized the role of internal developmental forces in the production of new species – views, such as those of Carl Nägeli, grounded in an earlier idealism – Darwinism offered a theory based upon the central role of the environment. Despite widespread interest in evolution theory, few nineteenth-century biologists actually made direct use of the concept of natural selection in their own research. The young German botanists under consideration in this study represent an important exception. Although Darwinism came under attack in Germany in the 1890s, and although much of the criticism originated in the botanical community, the growing interest in adaptation among younger German botanists in the last two decades of the nineteenth century nevertheless owed its inspiration, in large measure, to the theory of natural selection.

PLANT ADAPTATION IN GERMANY TO 1880

References to adaptation are almost entirely absent from the works of Hofmeister, von Mohl, Pringsheim, de Bary, and other early contributors to "scientific botany," as outlined by Schleiden. Whatever interest there may have been in adaptive phenomena was discouraged by the new botanical professionals who dominated the discipline from the 1850s through the 1870s. Determined to develop botany into an inductive science and rid it of any lingering speculative underpinnings, the new professionals emphasized description and cautious experimentation, and they discouraged any discussion of the purposes of organic structure. One of the effects of Schleiden's eloquent condemnation of speculative science was to make botanists think twice before discussing structure and function in the same breath, for fear that their work would be considered teleological. Writing in 1898 regarding this earlier state of affairs, Karl Goebel remarked that "for a long time it was nearly forbidden to call special attention to the 'purposefulness' [*Zweckmässigkeit*] of particular structural features of organisms."[1] Julius Wiesner, Haberlandt's mentor at the University of Vienna, concurred with Goebel, but he thought that a division of labor was necessary for a while in order to establish botanical science on a sound empirical footing:

> There came about a sharp sundering of morphology from the doctrine of function – so sharp that it was regarded as dangerous and punishable for one of these subjects to deal with things pertaining to the other. Under the chastisement of Schleiden no one attempted to demonstrate the functional significance of a morphological structure. Narrow-minded as this method of procedure appeared, it was to the purpose. Embryology of plant organs arose out of these conditions, and physiology was gathering richly of usable constructive material for the future.[2]

Wiesner was exaggerating Schleiden's personal role in this matter. Although the latter's *Grundzüge der wissenschaftlichen Botanik* set the stage for the inductive approach to botany that dominated the 1850s, 1860s, and 1870s, many of Schleiden's contemporaries shared his belief that plant science needed to enter a period of careful laboratory research and to set aside, for the time being, the larger questions that might lead only to philosophical dispute and scientific bankruptcy.

Schleiden and his contemporaries tended to treat teleological reasoning as one and the same thing, whether it was found in the argument from design of natural theology, in the speculative vitalistic doctrines of Lorenz Oken and other *Naturphilosophen,* or in the comparative anatomy of Cuvier. Yet Immanuel Kant, to whom Schleiden, the *Naturphilosophen,* and just about everyone else in German science traced their roots, had made a convincing argument for the inescapable need for a teleological perspective in the life sciences. In the *Critique of Judgement* (1790) Kant emphasized the limits of mechanistic explanation in accounting for organic phenomena:

> In a thing that we must judge as a natural purpose (an organized being), we can no doubt try all the known and yet to be discovered laws of mechanical production, and even hope to make good progress therewith, but we can never get rid of the call for a quite different ground of production for the possibility of such a product, viz. causality by means of purposes. Absolutely no human reason (in fact no finite reason like ours in quality, however much it may surpass it in degree) can hope to understand the production of even a blade of grass by mere mechanical causes. As regards the possibility of such an object, the teleological connection of causes and effects is quite indispensable for the judgment, even for studying it by the clue of experience.[3]

The purposiveness of living matter is thus the irreducible ground upon which a science of life must be constructed. Timothy Lenoir has described in detail the origin, nature, and fate of the research program that resulted from this insight in the early nineteenth century under the leadership of Johann Friedrich Blumenbach, Johann Christian Reil, and Carl Friedrich Kielmeyer. It is Lenoir's contention that in the hands of these "teleomechanists," as he has labeled them, teleological reasoning had a positive heuristic value. The teleomechanist program helped produce such contributions to the life sciences as the embryological studies of Johann Friedrich Meckel and Karl Ernst von Baer and the functional morphology of Rudolph Leuckart.[4]

Whatever value this program may have had, the mainstream of German scientific botany from mid-century on categorically condemned teleological language and avoided discussions of adaptation. Yet if the concept of adaptation found no place in the work of the new laboratory-oriented botanists, and if it was absent as well from the work of taxonomists and floristic plant geographers, it was at least a central underlying assumption of the physiognomic approach to plant geography begun by Humboldt. Humboldt's writing had a wide appeal in Germany, as elsewhere, and his essay on the physiognomy of plants appeared in several editions of his popular *Ansichten der Nature.*[5] Although this work, in a sense, lay outside the mainstream of scientific botany, it nevertheless received some attention from professionals, particularly August Grisebach and Anton Kerner. Both men assumed, with Humboldt, that the general forms of plants – grasses, heath plants, succulents, palms, deciduous trees, and so on depend directly upon climate. Grisebach took Humboldt's list of sixteen basic physiognomic plant types and expanded it to fifty-four. He also introduced the concept of the plant formation (*pflanzengeographische Formation*) – a collection of plants of similar physiognomic type forming a distinct feature of the landscape (e.g., tropical rain forest, savanna, tundra, steppe).[6] Anton Kerner (1831–98) employed Grisebach's concept of plant formation in his popular account of the patterns and cycles of vegetation in the Danube basin, *Das Pflanzenleben der Donauländer,* a book that had considerable influence on the young Gottlieb Haberlandt.[7] Kerner offered more detail than Humboldt regarding the nature of particular vegetation types, but, following Humboldt and Grisebach, he chose to emphasize the broad aspects of vegetation cover that strike the traveler as he moves from one climatic zone to another.

Kerner later became a proponent of Darwinism, but *Das Pflanzenleben der Donauländer,* although it appeared after the publication of the *Origin of Species,* made no mention of Darwin's work. Humboldt, not Darwin, was clearly Kerner's source of inspiration here. Many passages resemble Humboldt's writing in both style and content. Kerner, like Humboldt, conveys a sense of the abundance of nature and the rejuvenating forces of life, themes with roots in the romanticism of the early nineteenth century. Kerner came to his work with a more extensive knowledge of local vegetation than did Humboldt, but he kept within the Humboldt tradition, restricting his discussion to the gross morphological characteristics of plants in each climatic region and occasionally adding references to the local soil conditions and their effects on vegetation. The following passages typify the level of resolution found throughout the work:

> Since the loam which is formed from the clayey rocks of the calcareous Alps does not differ perceptibly from that developed from the crystalline rocks of the central Alps either in chemical or physical relations, it is not surprising that the meadow formations of the two regions are alike in their principal features. This is the explanation of the fact that in places in the calcareous Alps exactly the same plants and plant communities are found as those which people the slopes of the crystalline Alps.[8]

> The forests of the Hungarian lowlands, on the contrary, retain a wintry aspect long after the snow has melted. The earth is naked even on the most exposed places or covered only with dead leaves. The shrubbery which forms the understory of the forest, like the forest trees themselves, is almost wholly made up of species that leaf out late, budding forth at a time when the late frosts so common in the continental lowlands become less frequent.[9]

Absent from this work are the descriptions of intricate interrelationships between plant structure and function that characterize Kerner's later work and that of his younger German associates, such as Haberlandt, Volkens, and Schimper. Absent also are references to competition and struggle; *Das Pflanzenleben der Donauländer* conveys to the reader a strong sense of the harmony of the plant world with its physical surroundings.

Grisebach's more scholarly *Die Vegetation der Erde* merits special attention here, since it was the main work dealing with plant adaptation available during the 1870s to the botanists under consideration in this study, and since it also did not receive its inspiration from Darwin's evolution theory. Writing more than a decade after the appearance of the *Origin of Species,* Grisebach made ample references to Darwin, but he chose to take a neutral position toward the question of the origin of adaptations. He presented an atemporal picture of the relationship of vegetation to climate and soil, only occasionally pointing to an instance where a Darwinian explanation might possibly shed light on a particular pattern of plant distribution. Although admittedly an evolutionist, Grisebach preferred not to discuss the mechanism of evolution. He stated his central premise very simply in the preface: "The plant is the expression of the most diverse interactions of inorganic nature to which its development adapts."[10] Concerning competition, the struggle for existence, and natural selection, he had little to say. He believed that we can trace the present distribution patterns of plant species over the earth back to particular vegetation centers, but beyond that there is little that science can explain; the origin and behavior of these centers lie outside our understanding. He questioned the capacity of evolution theory to provide adequate explanations for many of the phenomena of plant distribution. For example, he thought that Darwin's theory could not easily account for the considerable variation in the number of endemic species on islands of similar character roughly equidistant from the nearest mainland:

> What can Darwin's theory do in cases where we see that under similar influences the results are yet so dissimilar and where geology also provides no explanation? It cannot be maintained that the islands richer in species have been in existence longer, since, indeed, in particular instances even the contrary can be demonstrated.[11]

To state Grisebach's objection more directly: If the species unique to particular islands have arisen, through variation and natural selection, from a few colonizing mainland species, how is it that islands with almost identical geophysical and climatic characteristics now exhibit such diversity in the number of endemic species? Darwinists, then as now, would have been willing to provide answers to this question (perhaps pointing out differences in the size of the islands, as well as local variations in oceanic and wind currents), but Grisebach would have seen no value in exploring such conjectures. He chose to remain on safer ground, describing the present plant formations in each region and pointing out the adaptations of plants to specific climatic and soil conditions.

His discussions of particular adaptive phenomena were a source of stimulation for botanists later in the century; Tschirch, Volkens, Stahl, and Schimper all made numerous references to *Die Vegetation der Erde* in their work. Relying upon a much less sophisticated knowledge of plant anatomy and physiology than that available to later botanists, Grisebach presented fairly detailed descriptions of the adaptations of plants in each climatic region and offered conjectures regarding the functional significance of particular structures. The following is part of his account of the general characteristics of vegetation in the steppes:

> Succulent plants, in whose tissue juices accumulate, thus enabling the plants to survive for long periods without inflow through the roots, inhabit the saline soil of the steppe region, and belong to the chenopod form [one of Grisebach's fifty-four physiognomic types]. It should be mentioned in general concerning succulents only that evaporation in these plants is restricted sometimes by means of an epidermal armor – through the strengthening effect of the sedimentation of a solid substance on the external surface of the upper skin – and sometimes as a result of the salt in their sap, which, due to its solubility, the plants take up easily from the soil. The basis for the latter phenomenon is that a salt solution evaporates more slowly than pure water, because the salt exerts a restraining force on the solvent.[12]

This passage, and many others like it, reveals Grisebach's effort to take the problem of adaptation well beyond the broad discussion of physiognomic types provided by Humboldt. In his own plant geography textbook, A. F. W. Schimper credited Grisebach with thus opening up a new ecological direction in the study of plant distribution, but he added that Grisebach approached the new field from an outmoded viewpoint.[13]

Grisebach did not hesitate to employ teleological language in his discussions of plant adaptation. The following remark from the section on steppe vegetation is not atypical: "The formation of thorns as well is based upon a plan of organization that strives to resist evaporation, in that it reduces the number and size of surface organs, and through restricted development, reduces water consumption by the plant."[14] Ernst Stahl later criticized this passage as speculation that fails to take into account the role of thorns as means of protection against animals.[15] Stahl believed that Grisebach placed too much emphasis on the adaptation of plants to climatic factors. That he discussed adaptation at all, however, set him apart from his professional colleagues. Whereas Hofmeister, Nägeli, Pringsheim, de Bary, and other botanists of their generation were busy at their microscopes, Grisebach was traveling extensively throughout Europe and Asia Minor. His travels alone, however, do not account for his interest in adaptation. Many German botanists made excursions in order to add to their collections but did not return home to write about the relationship of plants to their environments. In Grisebach's case, the intellectual climate of his university may have provided an additional source of stimulation.

Apart from a few years in Berlin, Grisebach spent most of his career, as both student and teacher, in the medical faculty at the University of Göttingen, where Blumenbach, who numbered Humboldt among his students, had established his Kantian program in biology. Following in this teleomechanist tradition, Grisebach's medical-zoological colleagues during the 1830s and 1840s – A. A. Berthold, Rudolph Wagner, and Carl Bergmann – held to a mechanistic, yet nonreductionist, view of the organism; they spoke freely in their work of purposiveness, in the sense of the necessarily close relationship between structure and function as the essential characteristic of living things.[16] Grisebach's interest in the study of plant adaptation dates back at least to 1838, when at the age of twenty-four he published his first work concerning the relationship between climate and plant distribution. His later plant geography text was an expanded version of the earlier paper, as Grisebach explained in the preface to the larger work.[17] Although published at a time when descriptive microscopic studies dominated the German botanical literature, *Die Vegetation der Erde,* with its emphasis on plant adaptation and its liberal use of teleological language, reflected the influence of the Göttingen school earlier in Grisebach's career.

If Grisebach's botanical colleagues in the mid-nineteenth century did not share his concern with plant adaptation, they shared his attitude toward evolution theory. Acceptance of the theory of descent but rejection of, or skepticism toward, natural selection was typical of the early response to Darwinism in Germany. Many German biologists considered natural selection too speculative a hypothesis to warrant serious attention. It did not fit in well with the laboratory emphasis of their work, it appeared too much like a revival of the speculative biology of the early part of the century, and it had no place among earlier German conceptions of evolution, most of them rooted in idealistic morphology. One group of critics of natural selection, which was to include eventually (and for different reasons) such prominent botanists as Carl Nägeli, Julius Sachs, and Karl Goebel, believed that Darwin's theory placed too much emphasis on adaptiveness and gave too great a role to the environment. Nägeli emerged early as the chief proponent of this view and continued to exert his influence well into the 1880s.

When in September 1885 August Weismann addressed the annual meeting of the Association of German Naturalists and Physicians on the subject of "The Significance of Sexual Reproduction in the Theory of Natural Selection," he used the occasion to attack Nägeli's recent reassertion of his belief that the guiding force in evolution is not natural selection but the innate tendency of organisms to produce variation toward greater perfectibility.[18]

Nägeli accorded a minor role to natural selection in weeding out clearly ill-adapted forms, but he believed that phylogenetic development is directed primarily from within by a "perfecting force" (*Vervollkommnungstrieb*) located in a hypothetical substance, the idioplasm. Weismann thought that Nägeli's emphasis on inner forces rather than on adaptation was due, in part, to his botanical perspective:

> I can fully understand how it is that a botanist has more inclination than a zoologist to take refuge in internal developmental forces. The relation of form to function, the adaptation of the organism to the internal and external conditions of life, is less prominent in plants than in animals; and it is even true that a large amount of observation and ingenuity is often necessary in order to make out any adaptation at all. The temptation to accept the view that everything depends upon internal directing causes is therefore all the greater.[19]

Weismann had a point. It is certainly easier to grasp at once the adaptive significance of sharp claws, prehensile tails, and thick fur than palmate leaves, red flowers, and parallel venation. He added, however, that in recent years botanists had begun to demonstrate the adaptive nature of many plant characteristics believed earlier to have only "morphological" significance, and stated that there was no reason to assume that researchers in the future would not eventually discover the functional role of all presently unintelligible structural features.

Although Weismann cited examples of work on plant adaptation carried out by botanists of his own generation, that is, those born in the 1830s and earlier, he seems to have been unaware of the growing interest in adaptation among younger botanists. At the time of his 1885 address, the early work of Haberlandt, Tschirch, Volkens, Stahl, and Schimper was already in print. This work may well have escaped Weismann's notice, but interest in adaptation had already surfaced by the 1870s in the work of some of his botanical contemporaries. Julius Sachs, inspired somewhat by Darwin, directed much of his physiological research toward problems involving the adaptive responses of plants to external stimuli; Hermann Müller, very much a Darwinist, had studied in detail the intricate adaptations of flowers to insect pollinators; and Simon Schwendener, although himself critical of natural selection, had, of course, published his study of the mechanical support system in monocots that served as a source of inspiration for his Darwinist students. In addition, Austrian plant physiologist Julius Wiesner, who may have exerted more influence on Haberlandt than the latter was willing to admit, published a study of the protective mechanisms of chlorophyll in plants; and Wiesner's eventual colleague at the University of Vienna, Anton Kerner, produced a work on the protective devices of flowers against uninvited insect visitors. Kerner, now influenced by Darwin and writing in a style that reminds us very little of *Das Pflanzenleben der Donauländer*, offered detailed descriptions of the intricate mechanisms by which plants are able to ward off potentially harmful intruders.[20] The literary style of the earlier book did return later in Kerner's most popular work, *Pflanzenleben*, in which he offered, for the layman, a fascinating and rather detailed account of a variety of plant adaptations, and in which he maintained his support for the theory of natural selection.[21]

PLANT ADAPTATION AND THE NEW DARWINISTS

After 1880 work on plant adaptation in Germany was dominated by the generation of botanists born around mid-century, men who had attended universities in the 1870s and early 1880s, when interest in Darwinism was at its peak in Germany.[22] Unlike their mentors, they received their botanical training entirely within a Darwinian context, in the sense that Darwin's evolution theory was an established, if still disputable, aspect of biological thought. Despite the hesitance of some of their teachers to accept natural selection, and despite the lingering pre-Darwinian interest in adaptation of the kind represented by Grisebach's work, the studies of adaptation initiated at Berlin, Bonn, and Jena in the 1880s found their justification and inspiration in the evolutionary mechanism proposed by Darwin. The works of these younger botanists emphasized the *struggle* for existence. In his work on epiphytes in the American tropics, for example, Schimper referred repeatedly to the "struggle for light" (*Kampf ums Licht*) as the determining factor in the evolutionary development of these plants; he was not content simply to describe the present suitability of various epiphytes to the different light conditions in the tropical forest.[23] Where Humboldt spoke of the harmonious relationship between plants and their environment, and Grisebach discussed vegetation almost as if it were the natural expression of the physical conditions in a given region, the new generation of German botanists emphasized the need for *protection* from harmful biotic and abiotic elements in the environment of the plant. We can list here Haberlandt's work on the various protective devices of seedlings, Stahl's study of the means of protection in plants against snails, Volkens's work on the protective mechanisms of plants against transpiration, and Stahl's study of the protective devices against rainfall in the leaves of tropical plants.[24] One can discuss means of protection in a non-Darwinian context, of course, but there is a clear shift of emphasis from Humboldt and Grisebach discussing the plant almost as part of the physical environment to the later botanists discussing the plant engaged in a continual struggle for survival. Even had they not repeatedly mentioned natural selection in their work, it would be difficult not to draw the conclusion that botanists in this latter group derived their inspiration from Darwin's theory.

In addition, these men were motivated by their commitment, with Darwin, to a gradualist interpretation of evolution, which assumed that all structures must have a functional significance in the life of the plant. Karl Semper, Julius

Sachs's zoologist colleague at Würzburg, stated this view quite well in his 1880 study of the conditions of existence to which animals must be adapted:

> . . . we perceive that most, perhaps all, of the characteristics not in a great measure hereditary originated through modifications of those originally adaptive organs which bore within them the elements of continuous and extensive gradual transformation.
>
> This inference includes another: That all the structural peculiarities of animals are true organs which must subserve some function and can never be mere useless ornaments. Otherwise, from the Darwinian point of view – which, as I have said, I accept as a standard – it would be quite unintelligible how wholly useless portions of the body could have been inherited and modified through a long series of divergent descendants from the present form.[25]

Semper admitted that there are peculiarities of structure that appear to be totally useless in the life of the individual, but he felt confident that future research would reveal a functional role for all such structures.

What was true for animals was also true for plants. Gottlieb Haberlandt stated that the central task of physiological plant anatomy is to "explain the anatomical characteristics of plant tissues on the basis of their physiological role. Accordingly, it applies Darwinian theory to histology and identifies the anatomical structure and arrangement of tissues as adaptive phenomena."[26] Haberlandt allowed for the possibility that truly useless structures exist in plants; and he explained these as structures that have lost their original function, have changed their function, or are closely associated with other clearly adaptive structures (as Darwin explained with his concept of the "correlation of growth"). Haberlandt did not believe that such cases pose a serious problem for Darwinian theory, and he assumed that these useless structures will eventually disappear, since natural selection forces upon organisms the principle of economy of material. "The struggle for existence compels the plant, where possible, to achieve the greatest effect with the least expense of material."[27] Haberlandt was clearly influenced by developments in energy physics since the mid-nineteenth century. For him, as for many Darwinists, the principle of the conservation of energy, in conjunction with the concept of natural selection, led to the inevitable conclusion that organisms that expend part of their limited energy supply on the production of useless structures could not compete successfully over a long period of time against more efficient rivals.[28] In later years, Haberlandt softened his position somewhat regarding the usefulness of plant structures, but in the third and fourth editions of his textbook he still asserted: "Even in its mildest form, however, the struggle for existence is active enough to prevent the perpetuation of a large proportion of useless morphological peculiarities in addition to the stock of useful adaptations with which every organism is provided."[29]

Haberlandt's general position was that one should assume that any given plant structure has a definite function in the living plant. Georg Volkens took this view to what he considered to be its logical conclusion. Assuming that all characteristics that distinguish one species from another must have adaptive significance, he stated confidently in the introduction to his study of Egyptian desert plants that in the future, by applying the physiological-anatomical method, a botanist will be able to develop a monograph of a genus that reflects the environmental conditions of the various habitats in which representatives of that genus are found.[30] In other words, the characters that he uses to distinguish different species will correspond very closely to the conditions in their respective habitats. Volkens believed that taxonomists must abandon their preoccupation with kinship and ask instead: "What causes the breakup of an originally monotypic species into numerous species? How have the different characteristics originated in the struggle for existence, and what is their relationship to present and past living conditions?"[31] Few, if any, of Volkens's colleagues would agree that the time was close at hand when plant taxonomists could differentiate species purely on the basis of adaptive characteristics, although Schimper made allusions to such a view at the conclusion of his study of strand plants.[32] However much this extreme selectionist evolutionary perspective entails, in principle, that all differentiating characteristics have a basis in environmental factors, in practice taxonomists would have to continue to differentiate species on the basis of purely morphological considerations, without reference to the adaptive significance of the characteristic in question.

If the time was not ripe for a fully reconstituted Darwinian taxonomy, these botanists at least agreed that Darwinism conferred a new respect on causal explanation in biology. "The time is gone," Alexander Tschirch declared, "when empty descriptions alone constitute the sciences, when morphological subtleties excite the spirit. Now we have advanced to the question 'Why?'."[33] Haberlandt stated that the proper subject of physiological plant anatomy is the causal nexus between morphological and physiological features that "is formed . . . during the development of an entire species and, indeed, very gradually, through adaptation by means of natural selection."[34] Further, Haberlandt believed that Darwinism brought the discussion of purpose back into prominence in biology. In the past, Haberlandt wrote in *Physiologische Pflanzenanatomie,* the teleological and mechanical modes of explanation were treated separately, and for good reason, but, he added, "it was left to the shrewdness of Darwin to find the mechanical formula for the teleological mode of explanation."[35] Many biologists, Ernst Haeckel and Julius Sachs among them, believed that Darwinism had spelled the end of teleology. However, their view was directed at the teleological notion that organisms, and their various parts, are designed for particular purposes.[36] Haberlandt, on the other hand, argued that although we cannot claim that organisms are designed for a purpose, purposiveness is the necessary consequence of natural selection, since the struggle for existence ensures the perpetuation of only those characteristics "that guarantee the safest, most complete, and most efficient operation of all physiological functions."[37] It is no longer necessary for biologists to feel that they are on safe ground only when they restrict themselves to the efficient (or mechanical) cause of a particular phenomenon. Darwinism has provided the link between final cause and efficient cause.

Haberlandt, Tschirch, Volkens, and the other physiological plant anatomists associated with Schwendener in Berlin did not produce treatises designed to demonstrate the workings of natural selection as the central causal agent in the production of the diverse array of plant structures observable at present. Their particular brand of functional anatomy assumed neither that the minutiae of plant structure have been designed for specific purposes nor that the close relationship between structure and function is simply the essence of the living condition and requires no further explanation. Instead, they assumed that the continual struggle for existence, and the inevitable effects of natural selection, leave no choice but that the details of organic structure serve specific functions, and as efficiently and effectively as possible. This was more than giving lip service to Darwinism. The previous generation of German botanists had ignored the question of adaptation in order to concentrate attention on developing a more complete understanding of embryology, morphology, anatomy, and physiology. Botanists of that generation were so used to the compartmentalization that had served them so well that they chastised Haberlandt early in his career for trying to deal with form and function together. He and his colleagues felt justified in so doing, however, because they believed that Darwinism dictated no other choice. Whatever valuable insights exclusive concentration on either structure or function had yielded in the past, they argued, the time had come to treat the two together. Not to do so would be to ignore the obvious explanatory power of the Darwinian synthesis.

Perhaps the best examples of application of the theory of natural selection to the interpretation of botanical phenomena occur in the works of Stahl, Schimper, and Schenck. These three not only assumed the primary role of natural selection in ensuring the intricate correlations between plant structure and function, they also actively pursued research problems in which they attempted to demonstrate the efficacy of natural selection in accounting for particular sets of adaptive phenomena. They fell far short of presenting full-blown expositions on Darwinian biology, but Stahl's work on plants and snails, as well as his investigation of leaf form in the tropics; Schimper's studies of tree–ant symbiosis and his various works on epiphytes and strand plants; and Schenck's elaborate study of lianas all indicate a strong commitment to a Darwinian interpretation of plant adaptation. Stahl, Schimper, and Schenck made a considerable effort to demonstrate the adaptive significance of specific plant structures, some of which may appear at first to have little functional value. They also took great care to point out that these adaptations, wonderful as they may be, did not simply arise spontaneously but came about in gradual, stepwise fashion as the result of completely comprehensible material causes. This interpretation allowed them to focus their attention on the environment of the plant, various elements of which serve as the selecting agents in each case.

THE CRITIQUE OF NATURAL SELECTION

Strong as it may have been among this small circle of ecophysiological adaptationists, the Darwinian view of plant adaptation was hardly the dominant view in the late nineteenth century. Resistance to a full-blown Darwinian interpretation of adaptation had always been present among German botanists, and by the last decade of the nineteenth century, as the general popularity of natural selection waned somewhat, this resistance reasserted itself. In addition to the persistent German approach to evolution that emphasized internal forces driving organic form toward phylogenetic development, the turn-of-the-century opposition to Darwinism also included a strong resurgence of interest in the inheritance of acquired characteristics.

When Carl Nägeli presented the latest version of his views on evolution in 1884, he repeated arguments that he had been using since the 1860s against a strict Darwinian interpretation of organic transformation. Nägeli did not deny that natural selection occurs, but he allowed it only secondary importance as a means of pruning poorly adapted varieties. He assigned the process of variation the central role in the production of new species:

> . . . according to Darwin, variation is the driving force, selection the directing and ordering force; according to my view, variation is both the driving and the directing force. According to Darwin, selection is necessary; without it a perfecting [*Vervollkommnung*] cannot occur, and related species would continue to exist in the selfsame state in which they were once found. According to my view, the concurrence [of closely related species] results merely in the elimination of the less viable; but it is entirely without influence on the achievement of greater perfectibility and better adaptability.[38]

In other words, more perfect, better-adapted organisms come into being because the force responsible for variation causes the production of more perfect varieties. Nägeli rejected the creative role of the environment, whether through natural selection or through the direct effects of external factors, in the sense of Lamarck, in heredity.[39]

In contrast to Nägeli's view, Stahl, Schimper, and Schenck, along with Haberlandt and the physiological plant anatomists, presented a strictly environmental explanation for the origin of adaptations, and they made clear their view that the role of the environment was indirect, through natural selection. None of them even seriously entertained the Lamarckian views, and only Schimper and Stahl took the time to point out their opposition to such views. Schimper mentioned at the conclusion of his study of Brazilian epiphytes that he took the position, with Weismann, that the environment does not exert a direct influence on the production of variation but acts only indirectly, through the "selection of the fittest variations at any given moment."[40] In "Regenfall und Blattgestalt" Stahl assured his readers that he did not believe that leaf forms well adapted to heavy rainfall have come about as the result of the direct effects of rain; and in "Pflanzen und Schnecken" he criticized Grisebach for suggesting that thorns and spines are direct adaptations to dryness.[41] Earlier works by Stahl, Schimper, or any of the others make no mention of Lamarckism; but this is hardly surprising, since neo-Lamarckism in Germany developed into a movement only in the late 1880s and 1890s as biologists began to take sides around the issue of Weismann's germ plasm theory. Until that time, many German biologists, including Weismann himself, did not rule out the possibility of the direct influence of environmental factors on the hereditary makeup of organisms. In the late 1880s, the initial impetus for neo-Lamarckism in Germany came from the zoologists, but a number of botanists became prominent figures in the movement as it developed.[42]

Stahl, Schimper, and the other botanists under consideration here maintained a strict selectionist viewpoint throughout the 1880s and 1890s. They were certainly not moved by Nägeli's attack on Darwinism. Whatever Nägeli's credentials as a cytologist, few botanists in the 1880s could take seriously the idea of an internal "perfecting force" directing the course of phylogeny. In the form in which Nägeli presented this concept, it smacked too much of an earlier idealism. However, the selection theory was not without its botanical critics. For the most part, the criticism was based upon morphological considerations, and even made use of some of Nägeli's arguments while avoiding his idealistic language. Before the neo-Lamarckian movement took hold, two of the most articulate critics of natural selection were Julius Sachs and Karl Goebel, both of whom were very much concerned with the problem of adaptation and the influence of the external environment upon plants. Sachs had included an eloquent account of the various external factors to which plants must become adapted not only in his botany textbook, but also in an equally popular series of plant physiology lectures first published in 1882.[43] Karl Goebel (1855–1932), Nägeli's successor at Munich, had studied with Hofmeister, de Bary, and Sachs. Through the influence of Sachs, he developed as his specialty the field of organography, the study of plant morphology from a physiological-developmental point of view, more or less the botanical equivalent of Wilhelm Roux's *Entwicklungsmechanik*. Goebel had been one of the earliest visitors to Treub's laboratory in Buitenzorg, and his two-volume *Pflanzenbiologische Schilderungen* contained many excellent studies of the adaptation of both European and exotic plants to a wide variety of environmental conditions.[44]

Both Sachs and Goebel had been sympathetic to Darwinism early in their careers, but both believed that Darwinists had taken the idea of natural selection too far. Sachs presented a faithful account of Darwin's evolution theory in his 1868 *Lehrbuch der Botanik,* but by the 1880s he publicly adopted a more neutral position. In *Lectures on the Physiology of Plants* he wrote:

> The theory of descent demands in the first place only the recognition of the fact that in the course of time organic forms have been produced by some concatenation of causes; it leaves to us, however, the responsibility of the answer to the question what forces have been effective in determining the production of organic forms in any given case.[45]

Sachs did not pursue the matter any further in that work; but by the 1890s he was willing to entertain the view that the process of variation, whatever its causes, plays a large role in phylogenetic development. He believed that natural selection could not account for the origin of major phylogenetic groups:

> Instead of fruitless attempts at explanation on the basis of selection, it is certainly better, and more consistent with the seriousness of the matter, and with the proven methods of scientific inquiry, to say *we do not know* how the major phylogenetic groups have originated and how from the simple initial forms they have developed into highly differentiated forms. [Sachs's emphasis][46]

Sachs thought that natural selection applies only in the narrowest of cases, those involving closely related species or varieties. For the differentiation of broader taxonomic categories, however, we must seek an explanation in the process of morphological development itself rather than in the influence of the environment. For example, Sachs considered the general increase in size and complexity of organisms within any phylogenetic series as a phenomenon lying entirely outside the domain of adaptation and natural selection.[47]

Some of Sachs's strongest criticisms of natural selection are to be found in two collections of notes published posthumously by his biographer, E. G. Pringsheim.[48] In these notes, written during an extended period of illness before his death in 1897, Sachs argued that specific adaptations may well be useful for an organism and yet have nothing whatever to do with the preservation of the species, "since specific adaptations are all the more deleterious for the distribution and propagation of a species the higher the degree of differentiation exhibited in its organs."[49] Sachs fully agreed with the view of the American paleontologist O. C. Marsh that the dinosaurs may have perished because of their too highly developed morphological differentiation. Regarding Marsh's view, Sachs stated, "this interpretation is a shorter expression of my own view that morphological differentiation advances entirely, or in part, independent of all adaptations and can even proceed to the detriment of species."[50] Sachs stated quite bluntly that the theory of natural selection "provides only the most incomplete basis for descent," since allied species are often distinguished by characteristics that have nothing whatever to do with adaptation and the struggle for existence.[51] Sachs attributed phylogenetic development to the effects of internal mechanical principles, about which we know very little at present and whose exact nature only a thoroughly physiological approach to plant morphology can reveal.

Sachs thus came to hold a view very similar to that of Nägeli (and the position that came to be known as "orthogenesis"). The essential difference between Sachs's view and that of Nägeli was that for Sachs the driving force in evolutionary change was an as yet unknown internal *mechanism* that drives the production of variation in a particular direction, whereas for Nägeli the agent of change was an internal impulse toward perfection. Like Nägeli, however, Sachs saw the effects of the environment as inconsequential, in the long run, beside the more powerful forces responsible for the production of variation. His student Karl Goebel, on the other hand, belonged to the generation of Haberlandt, Schimper, and Stahl. He criticized the view that natural selection is the creative force in evolution, but he attributed more significance than did Sachs to the role of the environment. In the 1890s Goebel presented these views in two brief but perceptive articles on the subject of plant adaptation. In the first, an article he prepared for the initial volume of the short-lived English journal *Science Progress,* he echoed Sachs's milder criticism:

> In the Darwinian theory of adaptation, organisation and adaptation are coincident, as the former arises from the gradual accumulation of useful alterations. But even on this explanation a series of organic conditions remains incomprehensible, especially the division of labour, which has really nothing to do with adaptation.[52]

Haberlandt had argued that the division of labor is highly adaptive, since it ensures the efficient and complete execution of physiological functions, and he regarded it as one of the central principles of physiological anatomy.[53] In support of his own view, Goebel described a genus of liverwort in some species of which the vegetative body forms a thallus and in others a leafy shoot. "It is inconceivable," he wrote, "why the leaf-formation should be more useful than the thallus-formation."[54] He believed that the plant world is full of such examples, whereby one plan of organization is just as good as another and the realization of the one or the other is unrelated to environmental factors. Goebel did not wish to discount the influence of the environment on plant form; his view was simply that in many cases environmental factors have little or nothing to do with the matter. As to the nature of the role of the environment, he indicated sympathies with Lamarckism: "Doubtless the theory of direct influence, which in a certain sense goes back to Lamarck, has much that is very attractive."[55]

In the second version of the article, which appeared four years later (1898), Goebel explored further the possibility of a Lamarckian interpretation of the origin of adaptations. He indicated that a number of features in plants – the thickness of the waxy cuticle of leaf surfaces, the development of fibrous tissue in response to mechanical stress, and overall stature and physical appearance – often vary considerably within the same species and seem to represent direct responses to the particular environmental conditions under which the plant is growing. It is not inconceivable, he argued, that some of these direct effects can become fixed over time within the hereditary makeup of the plant, since this seems to be the case for certain fungi and bacteria:

> There is no reason why we should not make the same assumption for the higher plants, that long lasting external influences can produce hereditary adaptation. It seems to me that the comparative investigation of adaptive phenomena within one and the same family forces such a conclusion, a conclusion which concurs with Lamarck's hypothesis that congenital characteristics can be acquired spontaneously.[56]

Goebel pointed out that Darwin himself did not believe natural selection to be the only cause of modification in living forms. "In reality," Goebel stated, "if we look around in the current botanical literature, we find that true Darwinism, that is, the tendency which assigns natural selection the central role in the production of adaptations, already has no more representatives in Germany."[57]

Goebel was overstating his case, but considerable opposition to the concept of natural selection had surfaced in Germany by the late 1890s. In 1888 zoologist Theodor Eimer attacked the strict selectionist position of Weismann on orthogenetic as well as Lamarckian grounds.[58] Others soon joined Eimer's attack on the Lamarckian side. By the turn of the century, neo-Lamarckism in Germany had developed into a movement led largely by younger men, such as zoologists Gustav Wolff and Ludwig Plate and botanists Richard von Wettstein and Adolf Wagner, all of whom were born during the 1860s.[59] Goebel's work was not really part of this new movement; he flirted with Lamarckism but did not pursue it as seriously as his younger colleagues. Since he had devoted his scientific career to investigating morphological development in plants, and since, like Sachs, he believed that many aspects of phylogeny do not lend themselves to a simple explanation based upon natural selection, he was attracted to the idea that the environment of the plant could exert a direct influence on the form of particular organs. Whereas Sachs thought that such effects are transitory and have little to do with phylogenetic development, Goebel was at least willing to entertain the possibility that the environment can produce more lasting changes.

Haberlandt, Tschirch, Volkens, Stahl, Schenck, and Schimper accepted from the start a primary creative role for the environment, not through its direct effects on the development of particular organs but through its selection of survivors from among the variant plant forms present in any given generation. As an experimental morphologist, Goebel had made comparative studies of the development of individual plant organs in numerous species. This embryological viewpoint convinced him that variation occurs only in particular directions as a response to as yet unknown internal forces. Not sharing Goebel's morphological-embryological perspective, the selectionists accepted variation as random and suggested mechanisms by which the selective activity of specific environmental factors may have produced complex plant adaptations by continually sifting out less well-adapted forms. Exclusive reliance upon natural selection allowed them to explain, or attempt to explain, all adaptive phenomena on the basis of known causes. They ignored any creative contribution from the process of variation itself and concentrated instead upon the selective effects of familiar environmental factors on the wide spectrum of variant plant forms that is normally present. In a sense, they were extending to the problem of adaptation the same parsimonious attitude that an earlier generation had restricted to the study of plant cells and tissues.

9

The colonial connection: imperialism and plant adaptation

Darwinism was not the only influence upon the environmental perspective of Haberlandt, Schimper, and their contemporaries. These botanists had traveled extensively and observed plants living under a variety of conditions very different from those found in the European environments with which they were familiar. Such experiences left a lasting impression on all who shared them. Visiting Germany's new colonial acquisitions, and those of other European nations, in the period before the German government attempted serious economic exploitation of the colonies, they tended to view their experiences, for the most part, from the point of view of pure scientific research. They used their opportunities in exotic environments to explore further the methods and theories that they had developed in their laboratories and institutes in Germany and had already begun to apply to the plant in its natural habitat. They treated the new colonial territories, as it were, like vast outdoor laboratories in which to extend the work begun in the indoor laboratories of Europe. They would have had few such opportunities, however, had it not been for the economic promise associated with the colonies.

Whether the science resulting from these overseas experiences was "imperialist" in nature is another matter, depending upon one's conception of imperialism. Certainly these studies of plant adaptation made very little use of the botanical knowledge of the indigenous peoples in the colonial regions, and they were hardly designed for the benefit of those peoples. The flow of ideas was from the German centers of learning to the colonies and back, with the overseas biota and environments serving only as sources of information to be fitted within a preexisting theoretical framework.[1] Some of the knowledge gained from this process was used to serve practical ends both at home and abroad, as in the acclimatization of exotic plants in Germany or in other colonial regions, but for the most part, the knowledge so gained was to serve the interests of German science. This was a subtle but no less one-sided form of exploitation that certainly fit within the broader framework of direct physical and economic subjugation usually associated with the colonial enterprise.

THE PRACTICAL SIDE OF COLONIAL BOTANY

In 1892, on the occasion of the seventy-fifth anniversary of the founding of the 'sLands Plantentuin, Melchior Treub received a glowing message of praise, printed on vellum, from ten German scientists who had conducted research at Buitenzorg. This tribute, signed by Haberlandt, Stahl, Schimper, and Goebel, among others, praised the garden for its rich botanical treasures and its excellent research facilities and asked that the "spirit of liberty, under whose protection alone a rich scientific life can develop, always find a home there for the benefit of the united scientific community."[2] Elsewhere, Schimper attributed the new physiological emphasis in plant geography in large measure to the research opportunities available at Buitenzorg; Tschirch considered the laboratory at Buitenzorg one of the few facilities in overseas lands capable of accommodating the new directions in botanical research; and Haberlandt stated bluntly that Treub's laboratory was the only facility in the tropics properly equipped for botanical research.[3]

These botanists recognized the value of the Buitenzorg Garden mainly as a facility for pure research. Treub, an academic at heart, shared their priorities. He had conceived of the garden and laboratory as a place for conducting pure research, and at first he showed very little interest in agricultural matters. Nevertheless, he had to justify the existence of the laboratory to the Netherlands government and to the Dutch planters in the Malay archipelago. Consequently, in addition to the laboratory for pure botanical research, he gradually set up a number of experiment stations, and he persuaded the planters to pay the salaries of assistants hired to conduct research on coffee, tea, tobacco, indigo, and other valuable crops. By the turn of the century, the botanical laboratory was just one part of a large biological and agronomic complex, whose staff had grown to include twenty-two scientific and technical personnel, most of them occupied with agricultural research.[4] Although his personal research interests and his sympathies remained on the side of pure science, Treub had to answer to the needs of a small island empire that played a rather important role in the Dutch economy. He knew who paid his bills, and he felt a sense of duty to the home government.

Treub's German visitors, generally speaking, did not share directly in his practical concerns; they came to Buitenzorg primarily to enjoy the opportunity to conduct pure research under the nearly ideal conditions available at the laboratory and at the mountain station in Tjibodas. Most of them were members of an academic elite that prided itself on its preoccupation with purely intellectual pursuits. For the most part, the botanical research carried out by these academics was unconnected to research and practice in agronomy and forestry. There was intense interest, nonetheless, in the agricultural exploitation of the German colonies, and proponents of colonial development recognized the need for persons with botanical training. Botanists with university backgrounds did not escape the attention of colonial entrepreneurs and special interest groups. Members of the Deutsche Kolonialgesellschaft – most of them manufacturers, merchants, civil officials, and military men – actively solicited the aid of trained personnel to foster their plans for colonial agricultural and economic development. They could, and usually did, draw on Germany's many agricultural and technical schools or on the half dozen or so university agriculture departments, but they were sometimes able to conscript the services of a graduate of one of the more prestigious university botanical departments. Such was the case, as we have seen, with Georg Volkens and Albrecht Zimmermann.

Zimmermann, frustrated with a succession of low-paying positions as an assistant or *Privatdozent* at various German universities, accepted an appointment at the Coffee Research Station in Buitenzorg, one of Treub's new agricultural facilities, in 1896. The station had been set up to revive Java's ailing coffee industry, and Zimmermann was hired as a botanical expert to assist the station's director, F. C. Kramers. He worked there for five years before taking a position at the Biologisch-Landwirtschaftliche Institut (Biological-Agricultural Institute) in Amani, East Africa, where the German colonial administration desired to undertake serious agricultural development. The institute, established in 1902, largely through the efforts and financial backing of the Deutsche Kolonialgesellschaft, conducted research on raising tropical plants, carried out agricultural experiments, disseminated information related to agricultural matters, and generally served as both an agricultural experiment station and a laboratory for applied research. Zimmermann, who became the institute's director in 1911, continued his coffee research there and added investigations into tropical forage plants, cotton, rubber, yams, bananas, and many other economically important tropical plants.[5]

If Zimmermann entered the field of colonial agriculture somewhat reluctantly at first, Georg Volkens did so as an enthusiastic proponent of colonial development. After his trip to East Africa in 1892–3, he became an active member of the Deutsche Kolonialgesellschaft and went on speaking tours to create interest in the economic expansion of Germany's colonial territories. In 1898 he refused the offer of an assistant professorship at Bonn in order to take a position at the Berlin Botanical Garden and Museum as curator of the Botanische Zentralstelle für die Kolonien. The idea for such an agency had begun to take form ten years earlier, when the governor of Cameroon wrote to Bismarck asking the chancellor to look into the matter of supplying the colony with seeds of economically valuable plants. Bismarck passed the governor's letter on to the Prussian Ministry of Culture, and that body, through the urging of the Deutsche Kolonialgesellschaft, eventually worked out a plan to set up a botanical office to serve the colonies. The Botanische Zentralstelle was formally established in 1891, when Adolf Engler made the final arrangements with the Colonial Division of the Foreign Office. The main function of the new agency was to assess the economic value of plant samples sent to it from the colonies and elsewhere and to explore the prospects for manufacturing marketable commodities from these plants. The Botanische Zentralstelle sent plant samples to chemical and drug companies for analysis and arranged to have plants sent to the colonies from all parts of the world to determine their economic potential in the new regions. Volkens's visit to the Buitenzorg Botanical Garden in 1901–2, for example, was made for the express purpose of procuring plant specimens to supply the botanical gardens and research stations scattered throughout the German colonies.[6]

The home of the Botanische Zentralstelle, the Berlin Botanical Garden and Museum, served as the central showcase for the plant treasures of the colonies and as a research institution for studying economically valuable plants. Smaller botanical gardens associated with other German universities served in that capacity as well; but the Berlin garden, by virtue of its size, its location, and the presence of the Zentralstelle, far overshadowed the other university gardens in its importance as a research facility for colonial botany. Concerning the role of German botanical gardens in general, Eduard Strasburger wrote in 1893: "As experimental and acclimatization gardens they are now called upon to advance colonial interests, and have as their model the magnificent achievements in which the Botanic Garden at Kew, near London, may justly take pride."[7] The Berlin garden was, in fact, modeled somewhat after Kew and served a similar role with respect to colonial agricultural development – cataloging the plant resources of the colonies, disseminating information regarding useful plants, conducting acclimatization experiments, and overseeing the activities of the colonial gardens.[8] Adolf Engler, who served as the director of the garden from 1889 to 1921, supervised the transfer of the facility to its new location at Dahlem, a move that required fifteen years (1895–1910) to complete. The new location provided more space for Engler's pet project, the arrangement of plants in natural groups corresponding to Grisebach's plant formations. It also provided more space for the cultivation, study, and display of useful tropical plants. Although Engler himself maintained his interest in pure plant systematics and plant geography, as director of the garden he made certain, through appointments of individuals like Volkens, that the interests of colonial botany were well served.[9]

In addition to botanical gardens and colonial research stations, a university-trained botanist might find employment at one of the institutes created to prepare planters, traders, farmers, and government officers for their overseas experiences, such as the Deutsche Kolonialschule at Witzenhausen, the Kolonialakademie at Halle, the Kolonialinstitut at Hamburg, or the Seminar für orientalische Sprache at the University of Berlin.[10] The Oriental Seminar, as this last-mentioned institute was commonly called, was established in 1887 to help create a corps of linguistically trained officers for foreign service. In addition to language instruction, the Seminar offered courses in geography, religions, and local custom.[11] Otto Warburg, the student of Anton de Bary who had reviewed the first edition of Haberlandt's *Physiologische Pflanzenanatomie,* became a lecturer in tropical botany at the Oriental Seminar in 1893 and remained there into the 1920s. Warburg (1859–1938), a native of Hamburg, traveled extensively after completing his doctorate and had been among the earliest German visitors to Treub's laboratory. A plant systematist by training, Warburg, like Volkens, was swept up in the German colonial movement. In 1897, he cofounded and coedited a new periodical, *Der Tropenpflanzer: Zeitschrift für tropische Landwirtschaft,* with Ferdinand Wholtmann, a professor of agriculture at the Royal Agricultural Academy at Bonn-Poppelsdorf. This periodical was the official organ of the Kolonial-Wirtschaftliches Komitee (Colonial Economic Committee), a separate organization that became affiliated with the Deutsche Kolonialgesellschaft in 1902. *Der Tropenpflanzer,* along with its supplement, *Beihefte zum Tropenpflanzer,* offered information and practical advice for the colonial planter, including reports on expeditions into interior regions, and economic reports on the progress of colonial agricultural enterprises.[12]

THE COLONIES AND PURE BOTANICAL RESEARCH

Volkens, Zimmermann, and Warburg represent exceptions to the general pattern. Despite the existence of the Botanische Zentralstelle and the use of university botanical gardens as experimental facilities for studying economically valuable plants, academic botanists did not play a large role in the exploitation of the economic potential of the German colonial territories, especially before the turn of the century. Yet they did not hesitate to exploit the research opportunities created by Germany's imperial venture. Plant systematists took advantage of the new colonial acquisitions to expand their collections and produce monographs on the flora of these regions. It comes as no surprise, for example, that Adolf Engler's major contributions to taxonomy and floristic plant geography concerned the vegetation of Africa. As director of the Berlin garden, Engler was in an excellent position to review the plant specimens brought in from expeditions in the new colonies and adjoining lands. Engler published scores of monographs on the African flora and eventually consolidated this work in his five-volume *Die Pflanzenwelt Afrikas.*[13] He was well aware that his systematic work might have practical applications, but he did not pursue such matters very far. Although he played an important role in the creation of the Botanische Zentralstelle, this agency did not become an active center for colonial botany until Volkens took over in 1898. Engler kept his colleagues and fellow citizens apprised of the activities of the Zentralstelle and of the general state of colonial botany, but he did not play a direct role in the practical economic affairs of the colonies.[14]

Academic botanists in Germany viewed the colonial regions as vast new experimental laboratories for testing theories developed at home. Here was a different kind of exploitation. Whereas patriots and entrepreneurs talked about the colonies as sources of raw materials for German industry and trade, the elite of the German botanical community saw in the colonial lands a rich source of raw materials for their intellectual pursuits. When Volkens traveled to Egypt in the 1880s, he did not go there to explore the agricultural potential of desert regions or to study economically important plants. Aside from his personal interest in travel and adventure, he went there because Schwendener had suggested to him that one can best study the relationship between plant form and function in regions with the harshest or most extreme climates. Volkens used the Egyptian desert vegetation to test theories regarding the nature of transpiration and photosynthesis. Similarly, he traveled to East Africa initially to study the effects of high altitude on the internal structures of plants.

Volkens was the most adventurous of Schwendener's students; he took quite seriously Schwendener's suggestion to study plant adaptation in regions with severe climates. Although the others did not attempt an ascent of Kilimanjaro, they at least made the trip to Buitenzorg. When they returned, their accounts of the experience always emphasized two themes. First, they wrote about the changing nature of botanical expeditions. If in the past one traveled to exotic lands to collect specimens of as many new plant species as possible, now one traveled with entirely different goals in mind. Alexander Tschirch wrote:

> One no longer sets out only to exploit uninvestigated regions or to gain more precise knowledge of every aspect of a previously explored region; now one goes with a particular scientific question in mind and sets about solving it, on the spot, by experimental means.[15]

Haberlandt devoted the introductory pages of *Eine botanische Tropenreise,* his popular work on the vegetation of Java and Malaysia, to a discussion of the new approach to botanical excursions. He echoed Tschirch's views, stating that the modern botanist travels to the tropics to carry out physiological experiments and to conduct biological (i.e., ecological) research.[16] Regarding the popularity of *Eine botanische Tropenreise,* he later wrote:

> This work owes its unusual and lasting success, for a travel book, to the fact that it undertook, for the first time, to treat the tropical plant world, especially that of Java, not from the viewpoint of systematics or plant geography but from the perspective of general botany. It was mainly the ecology, physiology, and organography of the tropical flora that came into its own in this book. I wrote it with a love and an enthusiasm that I devoted to no other work, and so it has always remained my favorite.[17]

The love and enthusiasm, we are to assume, were inspired as much by the ecophysiological perspective of the book as by its subject matter, that is, the vegetation of the East Indies. Since Haberlandt and Tschirch were well aware that systematics and floristic plant geography still played a large role in botanical excursions, their discussions of the new approach to these excursions were as much calls to action as descriptions of the current state of affairs. Similar calls to action can be found in the works of Stahl and Schimper.

The second theme emphasized by virtually all of these botanists was that one investigates tropical plants in order to gain a better understanding of native European vegetation and to dispel any misconceptions that may have come about as the result of a narrow, temperate-zone perspective. Thus Stahl wrote in his study of leaf form and rainfall in the tropics:

> The prevailing high atmospheric moisture content and the heavy and very regular precipitation are factors which must be accounted for in the organization of the plant. It is to be expected, therefore, that an investigation of tropical plants, directed at this problem, will shed light on many external and internal structural features, not only in tropical vegetation, which is adapted to an extreme climate, but also in the vegetation of the temperate zone, in which, for obvious reasons, adaptations to moisture are less prominent.[18]

Moisture is clearly more abundant in the tropics than in the temperate zone, but the tropical climate is not necessarily more "extreme" in every respect. The extremes of seasonal temperature variation are absent, and environmental conditions must be viewed as consistently favorable for plant growth and development. From such a perspective the seasonality of the temperate-zone climate appears as a severely limiting condition. Haberlandt offered such a view in *Eine botanische Tropenreise:*

> This completely universal adaptation to the long winter rest has impressed its stamp on all the plants in our native flora. We often forget this, because we have lived here all our lives. For similar reasons, we are seldom inclined to deduce general and far-reaching conclusions from other adaptations that relate only to the peculiarities of the European climate. In the tropical zone, on the other hand, where as a consequence of the uniformly warm and moist climate the external conditions for growth and nourishment are favorable throughout the entire year, the plant world can develop and grow with a freedom which our native flora is denied for the most part.[19]

Haberlandt stressed that in tropical plants all of the vital processes develop more completely and in a more typical fashion. He suggested that the tropical plant become the standard for studying vegetation in other regions, concluding that we must begin with an examination of the vital processes of tropical plants "if we want to understand and appreciate correctly the phenomena which the plants of our European homeland exhibit."[20]

There was an implied evolutionary argument in this view. In Haberlandt's time, as today, the tropics were considered the original home of flowering plants, and a tropical climate was known to have extended far north of its present boundaries well into the Tertiary period. In comparison to those of the tropics, the present plant communities of the north temperate zone are considerably less diverse. If the environmental conditions of the tropics are to be considered the norm, then the plants of the north temperate zone must be considered the hardy survivors whose distant relatives were able to withstand the unusual temperature and moisture regimes in that region. These plants can hardly be used as the standards upon which to base the study of adaptive phenomena; rather, they must be viewed as specialized exceptions that evolved from plants adapted to the more favorable conditions found in the tropics.[21]

A. F. W. Schimper began his study of coastal vegetation in the Indo-Malaysian tropics by describing the depauperate state of the vegetation on the European coasts: "The European strand supports only a meager vegetation. The ground is sparsely covered, the forms are not very diverse, trees are either lacking or stunted."[22] He contrasted this picture to the situation in the tropics, where coastal vegetation is lush and quite diverse. In this work, and in his earlier paper on the protective devices of coastal plants against transpiration, Schimper suggested that there was much to be learned about temperate-zone vegetation in general from studying tropical strand plants. In the earlier work, it will be recalled, he used his research on the coastal plants of Java to draw conclusions regarding the deciduous habit of trees and shrubs in the temperate zone.[23] In his plant geography textbook, Schimper stressed the need for botanists to study vegetation in all nontemperate climatic zones, not just the tropics:

> The greater prominence of physiology in geographical botany dates from the time when physiologists, who formerly worked in European laboratories only, began to study the vegetation of foreign countries in its native land. Europe, with its temperate climate and its vegetation greatly modified by cultivation, is less calculated to stimulate such observations; in moist tropical forests, in the Sahara, and in the tundras, the close connection between the character of the vegetation and the conditions of extreme climate is revealed by the most evident adaptations.[24]

Consequently, the transfer of the botanists out of their European laboratories and into the natural environment of the plant, wherever that might be, has greatly enhanced the study of plant adaptation in general and the particular physiological approach emphasized by Schimper and his colleagues.

Examples of this theme come up in the works of Volkens, Schenck, and Tschirch as well. Volkens, as mentioned, made use of desert plants in Egypt to test hypotheses developed in studies of native European vegetation. Heinrich Schenck, in his study of lianas, stated that the scarcity of woody climbing plants in central Europe suggests that only in the tropics, where one finds such plants in abundance and great taxonomic diversity, can one properly investigate their characteristic features.[25] Alexander Tschirch placed this theme within a Darwinian context, offering a different twist to the evolutionary argument for studying tropical, rather than temperate, vegetation. Since conditions in the tropics are quite favorable for plant growth, one should expect competition to be intense there as well:

> Where so many individuals exist beside one another in so small a space, and where new individuals are springing up all the time, only the right of the strongest prevails. The struggle for existence is nowhere more severe than here. If a plant sets roots in the soil, it strives to grow out of the semi-darkness of the understory and toward the light; and if it is able to grow or climb quickly, then it may attain its goal ahead of other plants, which are then prevented from doing so. The strongest, the most versatile, and the most ruthless are victorious; the modest are relentlessly crushed and destroyed – the resplendent forest exists upon innumerable corpses.[26]

Here, then, was the ultimate justification for a botanist of Tschirch's generation to study in the tropics. Not only is the vegetation more diverse and particular adaptations more prominent than in the temperate zone, but the struggle for existence is also more intense – as one should expect from the combination of general fecundity with limited space. Since the struggle for existence is the basis for the close relationship between plant structure and function, the tropics should provide the most appropriate setting for studying this relationship.

One finds in the early work of this group of German botanists very little discussion regarding the economic uses of plants found in the new colonial regions or the significance of the native vegetation for the local inhabitants. If they chose to remain aloof from the economic realities of imperialism, these academic botanists nevertheless engaged in a kind of intellectual imperialism. They saw the colonial lands as vast experimental areas for testing hypotheses developed at home, and they stressed repeatedly the application of knowledge gained in the tropics and other exotic regions in deciphering the mysteries of adaptive phenomena in native European vegetation. Their response to Germany's entry into the colonial arena was to take advantage of new opportunities to extend their field research and deepen their understanding of plant adaptation. Was the relationship between rainfall and leaf form difficult to grasp in the temperate zone, were the means of protection against animals less conspicuous in native European plants, were the protective devices against transpiration difficult to comprehend? Then one should investigate these phenomena in regions where they are more exaggerated and therefore easier to understand, where the relationship between environmental conditions and plant structures is more clearly defined.

These German botanists were as well prepared for this undertaking as any group of biologists in Europe. Whether there was a comparable group of zoologists is somewhat problematic. There were individuals, such as Karl

Semper, whose interests centered on animal adaptation; but one would be hard pressed to identify a group of zoologists, in Germany or elsewhere, with such consistent and extensive interests in adaptive phenomena as this group of German botanists. In the second half of the nineteenth century, zoologists were preoccupied with the study of phylogeny. The Darwinists at Jena led the way in such pursuits. Although he coined the word "ecology," Ernst Haeckel did not himself engage in ecological research, nor did his students. Richard and Oscar Hertwig were occupied with problems of invertebrate phylogenetic development and with related studies in embryology; Anton Dohrn, founder of the Zoological Station at Naples, busied himself with studies of arthropod phylogeny; and Arnold Lang, who studied with Dohrn at the Zoological Station, was concerned with the investigation of phylogenetic theories involving the formation of body cavities and the development of the annelids. Toward the end of the century, experimental studies in embryology occupied much of the attention of zoologists. Karl Möbius and Victor Hensen, both of whom studied the relationship between marine environments and populations of aquatic organisms in the Baltic Sea, represent two notable exceptions to this pattern; and I believe that it is in the work of marine biologists and limnologists that historians should look for further exceptions.[27]

If zoologists in general were not inclined to study adaptation, they also had more limited opportunities to carry on their research in exotic lands. Zoologists went along on expeditions to the colonial regions, but there was no zoological counterpart to the Buitenzorg Botanical Garden, where animal scientists could combine field studies with laboratory investigations. Botanists had such opportunities, and the young Germans under consideration here made good use of them. Well schooled in the principles of plant anatomy and physiology, inclined from the beginning to view form and function together, and stimulated by their strict adherence to Darwinian theory to search for the adaptive significance of every detail of plant structure, they could transfer their research programs directly to the exotic environments associated with the colonial regions. Their experiences in those environments, in turn, served to reinforce their already strong interest in adaptive phenomena and thus sustain and advance those research programs.

ACADEMIC EMPLOYMENT AND PLANT ADAPTATION STUDIES

The German university system also served to reinforce their research programs. Not only did the universities provide them with the professional training that allowed them to take advantage of opportunities to travel in foreign lands, but their long period of internship, which was a consequence of the scarcity of professorships, coincided for most of these botanists with the period in which Germany was acquiring a colonial empire, a period in which

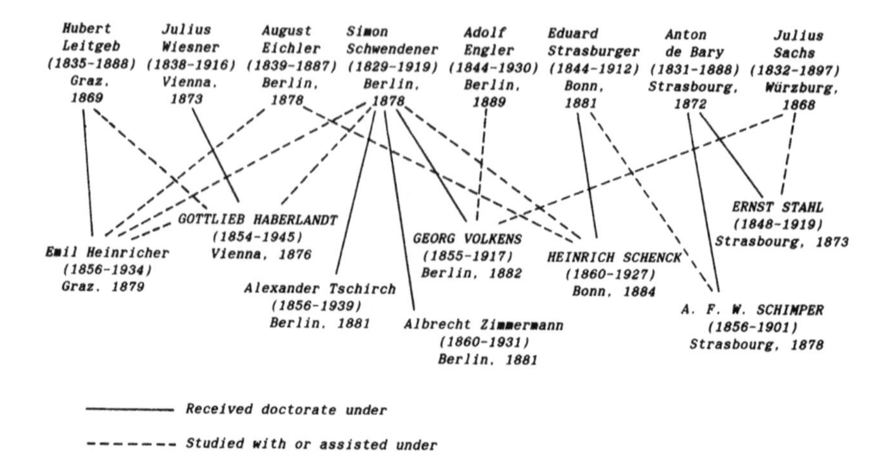

Institutional and intellectual lineages. Individuals treated in this volume (below), with institution and year of doctorate, and the botanists under whom they studied (above), with major institutional affiliation and year of appointment there.

economic exploitation had not begun on a large scale but in which there was general intellectual curiosity regarding the new colonial regions. These young botanists were not conspirators; they did not lobby for colonial expansion in order to expand the domain of their research interests. Nevertheless, they took advantage of opportunities to take their research far afield at a time when their careers were just beginning and their futures were still uncertain.

The botanists under consideration here – Haberlandt, Alexander Tschirch, Georg Volkens, Emil Heinricher, Albrecht Zimmermann, Ernst Stahl, A. F. W. Schimper, and Heinrich Schenck – were all born within a few years of each other. With the exception of Stahl (1848), they were all born between 1854 and 1860. In 1885, when their average age was thirty, all had received their doctorates, but only Stahl held a university chair. None of the others even held an assistant professorship, which in Germany usually carried little status or monetary reward. Haberlandt, Heinricher, and Tschirch were teaching at technical schools, Schimper was a *Privatdozent* at Bonn, and the others either were working as laboratory assistants or were unemployed. A decade later, Heinricher and Haberlandt, both natives of Austria, had found positions at Innsbruk and Graz, respectively; but among the remaining German-born botanists, only Tschirch held a university chair, and his appointment was in a Swiss university, and in pharmacology, not botany. Volkens, Zimmermann, Schenck, and Schimper were still looking for permanent university positions. It is not surprising that these four spent the most time in overseas territories. Zimmermann carried out only agricultural research in the colonies. Although he attempted some studies of plant adaptation at the Biological Institute in

Amani, his work in this field was confined, for the most part, to his early years in Berlin. The others, however, all carried on pure research in exotic lands, and some of their projects were quite extensive. Schimper made three trips to the American tropics and another to the East Indies before 1890, Volkens spent ten months in Egypt in 1884–5 and an equally long period in East Africa in 1892–3, and Schenck remained in Brazil for nearly a year in 1886–7. Those with regular university appointments could not spare as much time from their teaching and administrative responsibilities. As a new assistant professor at Bonn, even Schimper had to restrict the length of his trip to Brazil in 1886 in order to return to his teaching duties.

This is too small a group from which to draw firm generalizations, but Schimper, Schenck, and Volkens all produced book-length studies of various aspects of plant adaptation, whereas, among the others, only Stahl made substantial and consistent contributions to the ecologically oriented botanical literature. Although Haberlandt's trip to Java and Malaysia in 1891–2 led to two important studies of adaptation in tropical plants, he did not make additional trips, and his work at home remained confined to the laboratory. Tschirch made a brief visit to Buitenzorg in 1889 but then took on his new responsibilities as a professor of pharmacology and ceased research into plant adaptation altogether. Heinricher, who made only one two-month visit to Buitenzorg in 1903, produced several studies of parasitic plants but no important single work on plant adaptation. In general, aside from the early papers produced under Schwendener in Berlin, the major ecological studies carried out by this group of Darwinian botanists found their source of stimulation in field experiences in exotic lands. Even Stahl, who conducted a number of studies of plant adaptation in the vicinity of Jena, produced one of his major works as a result of his trip to Buitenzorg in 1889–90. Darwinism was an important source of stimulation for research into plant adaptation, but for most of these botanists it was not sufficient. The exploitation of this new field of research required experiences in novel environments, where particular adaptations stand out more clearly. Those who were in the best position to take advantage of such experiences were the botanists whose careers were still uncertain; they had less to lose and considerably more time on their hands.

10

Toward a science of plant ecology

In a tribute included as part of the English edition of *Pflanzengeographie auf physiologischer Grundlage,* British botanist Percy Groom credited A. F. W. Schimper with having founded the science of plant ecology. The subject barely existed, Groom stated, before Schimper began his work, and the few botanists who had attempted ecological studies employed inadequate and unscientific methods, "for the serious botanists were mainly working in their laboratories or in their herbaria." But Schimper changed all that, and not simply with his individual contributions to this new field:

> Far-reaching and highly original as Schimper's direct discoveries on oecological questions have been, botanical science owes to him a deeper debt for his foundation of a truly scientific and comprehensive method of oecological investigation resulting in the attraction of able botanists to work at this branch of the subject. Schimper from the first insisted on the employment of methods as strict as those used in solving morphological and physiological problems. And he showed himself the master of oecological method by his critical and concurrent use of three distinct modes of investigation, namely, of observations on the comparative morphology including histology, on the physiology, and on the geographical distribution of plants.[1]

In this passage Groom may as well have been describing the methodology of Volkens, Haberlandt, Schenck, or Stahl. Although Schimper developed this approach further than did any of the others, the close integration of knowledge of plant structure and function with intimate knowledge of the natural environmental conditions of the plant characterized the work of all of these Darwinian botanists.

From a late-twentieth-century perspective, however, Groom's insistence on Schimper's central role in the founding of plant ecology seems mistaken, or at least greatly exaggerated. Plant ecology came into prominence as a discipline in the early twentieth century as a result primarily of the efforts of botanists and plant geographers whose work focused on the plant *community* and its relationship to the environment.[2] Although Schimper devoted considerable attention in his textbook to plant formations and associations, his central concern was with building a science of ecological plant geography from the ground up by making use of elaborate studies of the relations of individual plants to their environments. This branch of ecology, termed "autecology" by Swiss botanist Carl Schröter, to distinguish it from the study of communities, or "synecology,"[3] did not provide the discipline-building impetus for ecological science. Most of the early plant ecologists took the view expressed by British ecologist Arthur Tansley in his 1911 work *Types of British Vegetation:*

> It may be said that we ought not to occupy ourselves with synecology till we have a complete or an approximately complete knowledge of autecology, but this is a mistaken notion. It might as reasonably be contended that we ought not to study the phenomena presented by the nations and races of men before we know all about the physiology and psychology of the individual man. As a matter of fact the study of synecology is considerably in advance of autecology (which is indeed still in a very backward state of development), and the progress made has amply justified the attention devoted to the wider though less fundamental branch of the subject. The plant-community, in fact, offers a convenient mode of approach to the study of plant-life in relation to the habitat. The systematic description and classification of vegetation affords a natural framework in which autecological studies will find their proper place.[4]

Tansley may have exaggerated the "backward" state of autecology, but he was certainly correct in identifying synecology as the driving force in plant ecology. Most of the work in this branch of the subject was done outside Germany, particularly in America, Scandinavia, and Britain.[5] Whether it was the study of the dynamics of plant community change over time (Cowles, Clements), the identification of stable community types by physiognomy or species composition (Flahault, Schröter, Du Rietz, Raunkiaer, Braun-Blanquet), or any of the numerous variations on these themes, the attention of these early plant ecologists (and phytosociologists) was always on plants in natural *groups*, with much discussion concerning the nature of the correct unit of vegetation study.[6]

Considerably more influential than Schimper's textbook in precipitating the founding of the new discipline was a book published in 1895 by Danish botanist Eugenius Warming. *Plantesamfund,* translated into German in 1896 as *Lehrbuch der ökologischen Pflanzengeographie,* was, for all practical purposes, the first textbook of plant ecology, and as the subtitle, *Ein Einführung in die Kenntnis der Pflanzenvereine,* indicates, the focus was on plants in association with each other.[7] Warming's textbook, especially the more accessible German edition, made an immediate impact on virtually all of the European and American botanists who were beginning to turn their attention to plant ecology. However, Schimper, whose own book was in preparation when Warming's first appeared, extended little credit to the Dane. In a report on the progress of plant geography from 1896 to 1898, Schimper mentioned Warming's work only in passing, commenting that it was "founded more on physiognomic than on physiological principles."[8] The emphasis of the book was clearly on assemblages of plants characterized by general form, but Warming, whose background included laboratory study in Germany, at Munich and Bonn, as well as extensive travel in the tropics, did not ignore the physiological dimension of plant adaptation studies. He made ample use of Schimper's own work, as well as that of Haberlandt, Heinricher, Schenck, Schwendener, Stahl, Tschirch, and Volkens.[9]

Nevertheless, it was the community emphasis, and not the references to autecological studies, that accounted for the immediate popularity of Warming's textbook as a source of inspiration for the development of a formal science of plant ecology. Warming's approach to his subject differed from that of Schimper and his colleagues in another important respect. Warming took a critical view of natural selection, placing equal emphasis, when he considered the mechanism of evolution at all, on the direct (Lamarckian) effects of the environment on plant form and structure. While recognizing the broad evolutionary context in which all biological investigations must be carried out, Warming nevertheless chose to center his ecological studies on the present set of relationships between assemblages of plants and their physical environments. This complex network of interactions between plant communities and environments in itself provided a well-defined subject matter for the new discipline; and by so defining that subject matter, Warming and the American, British, and continental European botanists who followed his lead also, in effect, removed the issue of the *origin* of adaptations from the realm of early plant ecology.[10] On the other hand, the strict Darwinism of Schimper, Stahl, and the other Germans considered here led them to focus their work on the relationship of the individual plant to its environment. Their selectionist evolutionary views, not to mention their strongly inductive, laboratory-oriented botanical training, led them to view the individual, and not the community, as the proper unit of ecological investigation.

The anatomical-physiological direction of ecological studies continued to have a following among German botanists throughout the twentieth century. In 1913 Franz Wilhelm Neger (1868–1923), a student of Karl Goebel, very much influenced by the works of Ernst Stahl, published a comprehensive plant ecology textbook under the title *Biologie der Pflanzen*.[11] Although Neger's book did not receive wide attention, it incorporated the work of all of the German botanists considered in this study and essentially continued in the same direction. Haberlandt's *Physiologische Pflanzenanatomie*, as mentioned earlier, appeared in several editions up to 1924, and Schimper's *Pflanzengeographie*, updated by F. C. von Faber, appeared as late as 1935. Hans Fitting (1877–1970), Strasburger's successor at Bonn, produced several ecological works along the line of Schimper's physiological plant geography.[12] The eventual successor to the 1935 Schimper–Faber book was Heinrich Walter's 1962 work (dedicated to the memory of Schimper), *Die Vegetation der Erde in ökologische Betrachtung.*[13] Walter (b. 1898) had been asked to revise the Schimper–Faber work in 1957, but he chose to write his own textbook instead. He began his ecological studies under Ernst Stahl at Jena in 1918, and his writings in ecological plant geography date to the 1920s.[14] In a sense, Walter's work constitutes a direct link between contemporary German plant ecology and the work of Haberlandt, Schimper, Stahl, Volkens, and their colleagues. Walter always made clear in his work the distinction between the study of plant communities, which he designated "plant sociology" or "vegetation research," and plant ecology, whose domain he restricted to the study of the physiological relations of individual plants to the physical limitations of particular habitats.

If the strong community emphasis of formal plant ecology, particularly as it developed in the United States and Great Britain, seems, in retrospect, to have emerged almost in disregard of this German work on plant adaptation, there were nevertheless numerous early ties between the late-nineteenth-century German ecophysiological studies and the founders of community-oriented plant ecology. To begin with, discussions over which term to use for the new discipline, a seemingly insignificant matter on the surface, reveal the links between ecology and the physiological direction in German botanical research in the late nineteenth century. As mentioned earlier, many Germans used the adjective "biological" to refer to works that we would now designate as ecological. By the last decade of the nineteenth century, the German *Biologie* could almost be translated directly as "ecology." To some extent, the ecological connotations of *Biologie* restored to that term the dynamic sense of a total science of life that the coiners of the word "biology" (Treviranus and Lamarck) had originally intended in the first years of the nineteenth century.[15] In any event, outside Germany "biology" had already begun to take on something like its present meaning, that is, the total subject matter of zoology, botany, and all related fields; and discussions had begun over whether to adopt Haeckel's term *Oecologie,* which he had originally coined in 1866, to apply to the branch of biology dealing with the relationships between organisms and their environments. Despite Franz Neger's use of *Biologie* in the title of his ecological textbook as late as 1913, German authors, including Haberlandt and Schimper, had begun substituting *Oekologie* for *Biologie* during the 1890s and early 1900s.[16] Discussions over which term to use were always framed within the context of physiology. Haeckel had originally referred to *Oecologie* as a branch of physiology; and when "ecology" was officially adopted in the United States in 1893, at a meeting devoted to botanical nomenclature, the issue came up in a discussion of the terminology of plant physiology.[17]

This connection between physiology and ecology persisted in American and British plant ecology. In 1908, Henry Chandler Cowles, one of the central figures in the development of American plant ecology, stated, "It is coming to be realized that the problems of physiology and ecology are identical." Frederic E. Clements, the reigning dean of American plant ecology for the first three decades of the twentieth century, stated, with his collaborator Roscoe Pound, in their 1898 study of Nebraska vegetation: "Ecology can not be set off sharply from physiology. Indeed, it is simply that particular phase of physiology which is manifested in the structure and habits of plants in their various homes." In his 1911 study of British vegetation, plant ecologist Arthur Tansley stated: "Ecology includes more than the study of vegetation-units or plant-communities; it deals with the whole of the relations of individual plants to their habitats. This latter branch evidently cannot be sharply separated from physiology; and may in fact be justly considered as a part of that subject."[18] American botanist V. M. Spalding, president of the recently formed Society for Plant Morphology and Physiology, chose as the topic of his 1902 presidential address "The Rise and Progress of Ecology." The following year the new president, William F. Ganong, spoke on "The Cardinal Principles of Ecology."[19] This choice of topics at two consecutive gatherings of plant morphologists and physiologists suggests that, at this early stage, plant ecology was certainly not identified solely with the study of plant communities. Indeed, both of these speakers, as well as Cowles, Clements, and Tansley, made ample references to the ecological work of German anatomical-physiological investigators as crucial to their new science.

In 1907 Frederic Clements, who taught plant physiology for several years at the University of Nebraska after completing his doctorate there, reissued his 1905 *Research Methods in Ecology* under a new title, *Plant Physiology and Ecology*. He maintained the connection between physiology and ecology and indicated that he was using the term "physiology" perhaps in a broader sense than it is usually used today:

> The proper task of physiology is the study of the external factors of the environment or habitat in which the plant lives and of the activities and structures which these factors call forth. The former are causes, the latter are effects. . . . Physiology was originally understood to be an inquiry into the origin and nature of plants. This is the view that pervades the following pages, and in accordance with this the subject-matter of ecology is merged with that of physiology.[20]

None of the Germans discussed in the present volume would have disagreed with that statement. Despite Clements's identification as the originator of the concept of the climax formation (the stable and integral vegetational assemblage associated with a particular climatic region), and despite his usual association with the German floristic-physiognomic plant geographer Oscar Drude,[21] Clements recognized a debt early on in his career to the German botanists who had concentrated their attention upon detailed anatomical and physiological studies of plant adaptation. He frequently cited their works, and he continued to maintain, in principle if not in practice, the essential identity of plant physiology and plant ecology.[22]

Connections between Henry Cowles and the Germans discussed in this volume are even clearer. Justly famous for his seminal studies of the succession of plant communities along the sand dunes bordering Lake Michigan,[23] Cowles was also the first American botanist to secure a position primarily as a plant ecologist. At the University of Chicago he was in an advantageous position to introduce the new field of study to an entire generation of American botanists and zoologists. His long teaching career at Chicago began in 1896, when he assisted his mentor, John Merle Coulter, in the first plant ecology course offered at the university. Two years later, upon completing his doctorate, Cowles was hired to teach and direct research in the new discipline. He remained in that position until his retirement in 1934.[24] In the summer of 1898, Cowles offered his first independent course. This was not "General Plant Ecology" or "Ecological Plant Geography" but "Ecological Anatomy." The description read as follows:

> Botany 23. Ecological Anatomy – This course presents the various plant structures from the standpoint of function. The tissue systems of plants are studied in succession, especial emphasis being laid upon environmental adaptations. The work is based mainly on Haberlandt's *Physiologische Pflanzenanatomie*.[25]

Although various other botanists taught the general plant ecology course at Chicago, as well as an array of more specialized ecology courses, Cowles himself continued to offer "Ecological Anatomy," with the identical course description, every year of his tenure at Chicago into the 1930s.[26]

An examination of the only plant ecology textbook that Cowles authored indicates that his preoccupation with Haberlandt's approach to plant adaptation was by no means confined to a single course. In 1911 he published the ecology part of a three-part botany textbook co-authored with his Chicago colleagues John M. Coulter and Charles R. Barnes. Coulter and Barnes wrote on plant morphology and plant physiology, respectively, in the first volume, and Cowles devoted the entire second volume to the subject of plant ecology.[27] He included a chapter on each of the following, in order: roots and rhizoids, leaves, stems, saprophytism and symbiosis, reproduction and dispersal, germination, plant associations, and adaptation. The chapter on plant associations covered 8 pages of a nearly 500-page textbook. Throughout this book, Cowles cited repeatedly the works of Haberlandt, Schimper, Heinricher, and Tschirch, as well as those of Fitting and Neger. Whatever the extent of his interest in plant communities, Cowles clearly saw the need to introduce students to plant ecology at the level of the relationship of the individual plant to its environment.

Cowles's and Clements's recognition of the value of the work of their German predecessors can be viewed as an acknowledgment of the need for autecological studies as building blocks for the more synthetic work of community ecology, but I believe that it has a broader significance for the history of plant ecology. When Charles Elton defined ecology as "scientific natural history," he expressed a simple and meaningful truth.[28] Toward the end of the nineteenth century, there was little room for the naturalist within the growing ranks of professional biologists. As all branches of life science became more "scientific," natural history, as such, lost its attraction for the serious scientist, and many aspects of natural history became incorporated within the framework of existing sciences and even led to the formation of new disciplines and subdisciplines.[29] Nevertheless, ecology is a field-oriented science that tends to attract biologists with a strong interest in natural history. Much of the ecological literature of the first half of the twentieth century had the character of natural history, except that it was somewhat duller to read. As scientific professionals and practitioners of a legitimate discipline, however, ecologists have always been sensitive to claims that their subject is little more than rehashed natural history, and they have taken pains to point out its rigor (or potential rigor) and its value as an explanatory science. There is a long literature in British and American plant ecology in which individual ecologists reflect on the true nature of their discipline, criticize the descriptive and shallow character of much of the work, and implore their colleagues to clarify their language, seek the underlying causes of ecological phenomena, and make use, wherever possible, of rigorous experimentation.[30] Whether or not plant ecology has always lived up to the original intent of its founders as a causal-explanatory science, a succession of plant ecologists over the years has maintained a continuous vigil to keep the science on the correct course. I believe that it is in this self-conscious effort to preserve the scientific character of plant ecology that we can locate a deeper connection to the earlier German work.

The appearance of a science of plant ecology in the early twentieth century was the result of the integration of several separate but not unrelated movements: the increasing attention to the habitat of the plant (encouraged by a variety of factors, from Darwinism to the sophistication of plant physiology); the growing interest in the geographical distribution of types, rather than species, of plants; the professionalization of botany; and the emphasis on functional explanation in the biological sciences.[31] All of these movements played a role in the development of German studies of plant adaptation in the late nineteenth century. Like the early American and British plant ecologists, Haberlandt, Stahl, Schimper, Volkens, and the other German plant adaptationists were professional botanists with rigorous training and a desire to maintain professional standards; but they were also field-oriented scientists with strong naturalist tendencies. Haberlandt's *Eine botanische Tropenreise,* Volkens's *Der Kilimandscharo,* and Tschirch's "Der javanische Urwald" are as much contributions to natural history as to scientific botany. Having naturalist inclinations from the start, and having prepared for their professional careers at a time when Darwinism was receiving serious attention in Germany, these botanists found in the concept of natural selection the theoretical justification for applying their laboratory training in anatomy and physiology to field-related problems. Within the very broad framework of late-nineteenth-century plant geography, they isolated specific topics regarding the functional significance of particular structural features of plants within particular environments.

The links between early community-oriented formal plant ecology and these late-nineteenth-century German studies of plant adaptation were by no means accidental. There were two senses in which the community ecologists needed to acknowledge a debt to, and maintain a connection with, the earlier ecophysiological studies. First, at the turn of the century, Germany was still recognized as the center for laboratory science; the detailed anatomical-physiological studies of Haberlandt, Stahl, Volkens, and their colleagues carried with them the authority and soundness of a well-established laboratory tradition. Over and above the immediate usefulness of these German studies for shedding light on particular ecological questions, they conferred a scientific legitimacy to a new discipline seeking to justify its autonomy. The young professional botanists who were forming the first generation of professional ecologists wished to align themselves and their work with the best of the German tradition in scientific botany.[32] The second sense in which the community ecologists needed to retain a link to the German anatomical-physiological tradition is a bit more indirect but no less important. From the beginning, community ecologists, especially in the United States and Britain, sought to do more than merely name and classify biotic communities. Their goal was to develop a dynamic science of community change over time. Frederic Clements may have been justly criticized for proliferating a cumbersome terminology of climaxes, postclimaxes, disclimaxes, and what not, but his ultimate goal was to discern the correct pattern of ecological succession associated with each climatic region. The physiological, that is, functional, approach to plant structure that characterized the work of the Germans treated in this volume was applied by community ecologists to whole assemblages of plants and eventually to entire biotic communities. The particular form of the interactions attributed to such communities may have depended upon the biases of the scientist, but the important element of this transfer of ideas was the adoption of a functional approach to community organization. From this perspective, both Clements's organic concept of the climax formation and Tansley's mechanistic concept of the ecosystem (introduced in 1935) can be viewed as having incorporated the late-nineteenth-century physiological ideal at a higher level of organization than the individual, producing an ecological science whose focus, at least in principle, was the dynamic set of interactions among biotic and abiotic components of natural systems.[33]

Cultural, institutional, and environmental factors may well have influenced the first appearance of a disciplinary structure for plant ecology in the midwestern United States. The expanding, flexible American university system could more easily absorb specialists in new areas of research. Although few vacancies were created specifically for plant ecologists, American universities could nonetheless add a specialist in plant ecology to an expanding botany department, and nowhere were botany departments expanding more rapidly than in the American Midwest. In Germany, by contrast, the limitations on the number of professorships continued well into the twentieth century, and this situation was only exacerbated by the devastating cultural and economic effects of the First World War. The form of theoretical plant ecology in America, the emphasis on the dynamics of vegetation change, was also tied to the experiences of botanists in the changing environments of the nation's heartland. The two early nuclei of self-conscious ecology, the universities of Chicago and Nebraska, were located in regions where remaining tracts of natural vegetation existed near rapidly advancing fronts of urban, industrial, and agricultural development. Change was a condition of existence in those regions where change became incorporated into the theoretical structure of the new science of plant ecology. In Germany, on the other hand, the existing "natural" vegetation represented a patchy mosaic of plant cover modified by centuries of European civilization. Whereas American botanists had direct access to tracts of pristine vegetation right outside their doors (although not for long), Germans could find such pristine conditions only in their travels to the tropics; and the rather complex, taxonomically diverse, and relatively stable plant communities they found in tropical environments would hardly have suggested to them simple patterns of successional change.

All of these factors suggest that in the United States the combination of institutional support with the creation of the first powerful ecological paradigm, the vegetational climax as an organic whole, led to the formal development of plant ecology in a manner that was not possible in Germany at the turn of the century. Additional support for ecological research in America came from the Carnegie Institution, which established a Desert Botanical Laboratory in Arizona in 1903 and offered a full-time position in plant ecology to Frederic Clements in 1916, eventually funding his research stations in the Colorado mountains and in Santa Barbara, California.[34] Here, then, was the leading theorist in American ecology fully supported by one of the largest private funding agencies. By 1920 the United States had an ecological society and a journal of ecology. Only Great Britain could boast a similar institutional structure for ecology.[35] Although British plant ecology owed its origins to regional and national surveys of vegetation initiated at the turn of the century,

British ecologists nevertheless acknowledged the influential role played by Clements's theoretical work in the establishment of the discipline in their country and the preeminent place of Americans in the new discipline.[36]

Recent studies of the development of ecological science in Russia and Sweden reinforce these speculations regarding the origins of formal plant ecology. Russian phytosociologists developed organismic conceptions of the plant community, similar in some respects to Clements's climax, during the 1890s and early 1900s. In his recent study of community ecology and conservation during the Stalin years, Douglas Weiner points out that by the 1920s ecology was finding some institutional support in Russian universities; and the organismic notion was extended from the plant community to the whole biotic community, or "biocenosis." By the early 1930s, however, this movement was effectively derailed by a confrontation with idealogues in the Stalin regime over the uses of nature reserves and over the appropriateness of a science grounded in the study of the long-term stability of "natural" interrelationships. The result was that those ecologists who did not change their careers were forced to redirect the focus of their research away from studies of the structure of natural systems.[37]

In Sweden, according to Thomas Söderqvist, plant ecology found little consistent institutional support until well into the twentieth century, owing, in part, to the rather limited number of university chairs in botany but also to a disagreement over the proper methodology with which to study plant–environment relationships. During the 1920s, proponents of two competing approaches to plant ecology vied for existing university appointments – a rather static sociological approach, based upon the identification of communities by their dominant species, and a physiological approach influenced by late-nineteenth-century German botany. The appointment of plant sociologist Einar Du Rietz to a key post at Uppsala in 1934 effectively dissolved the physiological school, led by Hendrick Lundegårdh, who had studied plant physiology in Leipzig under Sachs's student Wilhelm Pfeffer. However, Du Rietz's triumph did not lead to the establishment of a community orientation in ecology. Neither Du Rietz nor his followers regarded their sociological approach as falling within the province of ecology. Du Rietz identified ecology with attempts to discover functional relationships between the habitat and its plant cover, whereas plant sociology had as its aim only the identification of natural plant assemblages, a task that Du Rietz believed was difficult enough in itself but at least capable of being achieved. Since Du Rietz had developed his sociological approach in direct opposition to ecological plant geography and Darwinian studies of plant adaptation, Söderqvist characterized the effect of the Du Rietz appointment at Uppsala as a "de-ecologization of academia" in Sweden.[38]

In the United States, plant ecology was hardly a thriving discipline during the first half of the twentieth century; but due to a variety of institutional and intellectual factors, the new science was perhaps more visible there than anywhere else. It would be mistaken to associate plant ecology only with the American emphasis on the dynamics of vegetation change and the Clementsian organic metaphor for the plant community, but it would also be mistaken to underestimate the power of that metaphor as a means of identifying plant ecology with a legitimate piece of scientific turf at just the time when conditions were right for encouraging institutional support. The particular combination of institutional support and forceful scientific paradigm may not have existed in any other country during the early decades of the twentieth century. There are many questions yet to be answered regarding the establishment of plant ecology, and ecology in general, as a scientific specialization, but conditions certainly seem to have been more favorable for the new science in the United States than anywhere else.

Nineteenth-century German ecophysiological botany, which influenced the development of plant ecology in America, Britain, Sweden, and elsewhere, must be viewed not merely as an antecedent to these later developments but as a movement grounded in its own peculiar set of intellectual concerns and institutional constraints. For the individuals involved, it represented a completion and a fulfillment of the full potential of botanical science. Here was a rigorous approach to the study of plant structure that relied upon the experiences of the field researcher and the methodology of the physiologist, as well as the techniques of the microscopist. Observations and experiments were to be undertaken within the framework of a thoroughly functional approach to organic structure legitimized by a strict interpretation of Darwinian evolution theory. Haberlandt, Volkens, Stahl, and Schimper viewed their studies of plant adaptation as the ultimate expressions of the program of "scientific botany" initiated by M. J. Schleiden a half-century earlier. But whereas Schleiden had found fault with what he considered a dilettantish approach to field botany and had implored his colleagues to get to work in their laboratories and behind their microscopes, these young practitioners of plant science at the end of the nineteenth century, in many respects Schleiden's direct heirs, bestowed a new legitimacy to field research by incorporating laboratory techniques and attitudes into the study of adaptation.

Although the German colonial enterprise helped to reinforce this research school by making possible opportunities for travel in exotic environments, peculiarities of the German university system and the personal trajectories of the key individuals involved prevented the school from gaining a strong foothold in any one botanical department of a major university. Schimper, who never secured a permanent post in a German university, died young; Volkens and Zimmermann turned to colonial botany; Schenck, like Volkens, Zimmermann, and Schimper, had difficulty finding a regular university appointment; Stahl, aside from his possible influence on Heinrich Walter, hardly built up the little University of Jena into a thriving center for plant adaptation research; and Haberlandt assumed his post in Berlin long after his youthful enthusiasm for physiological plant anatomy had waned.

Also, by this time (1910), Haberlandt had lost his earlier zeal for a strict Darwinian approach to adaptation. In later editions of *Physiologische Pflanzenanatomie* he did not stress as much the Darwinian basis for his adaptationist views, although he continued to insist that most plant structures must be viewed as purposeful. In the fourth (1909) edition of his textbook he stated: "The recognition of the existence of adaptations in the internal structures of plants is completely independent of the various interpretations and hypotheses which seek to explain the origin of these adaptations."[39] By the first decade of the twentieth century, natural selection theory was under attack on many fronts in Germany and elsewhere. This was, after all, the period of the "eclipse of Darwinism," as Julian Huxley, and more recently Peter Bowler, have pointed out.[40] Although Haberlandt and his contemporaries may not have agreed with Goebel that Darwinism no longer had any representatives in Germany or, with von Dennert, that it was "on its deathbed," it was certainly true that varieties of neo-Lamarckism, orthogenesis, and saltation theory (i.e., De Vries's theory of macromutations) were getting considerable attention in Germany.[41] A program of research based on finding an adaptive significance for every structure, justified by a strict interpretation of natural selection theory, was going to find itself at odds with the growing anti-Darwinian sentiment in the early twentieth century. By undermining its primary rationale, this general critique of selection theory may have served to take the steam out of a research program that was already having difficulties finding a permanent institutional home.

All these speculations aside, the plant adaptation studies of Haberlandt, Volkens, Schimper, Stahl, and their colleagues were not ultimately rejected; they were absorbed within the existing structure of botanical teaching and research in Germany and abroad. Although it might be tempting to view their research program as failed or exhausted by the first decade of the twentieth century, the more accurate interpretation is that their work contributed to an increasing emphasis on the functional relationship between the plant and its natural habitat. Whereas in America and Britain, and to some extent in Scandinavia and Russia, this emphasis led to the establishment of a formal science of plant ecology, in Germany it led mainly to the continued application of physiological knowledge and principles to plant anatomy and plant geography. All these examples involve a synthesis of the laboratory tradition with the natural history tradition in botany. The late-nineteenth-century German studies of plant adaptation represent one of the first successful efforts to achieve that synthesis, to transfer the laboratory enterprise out into nature and at the same time to view nature as a living laboratory.

NOTES

Introduction

1. Goebel, *Ueber Studium und Auffassung der Anpassungsercheinungen bei Pflanzen* (Munich: K. B. Akademie, G. Franz, 1898), pp. 4–9.
2. Since the early twentieth century, ecologists have used the term "autecology" to refer to the study of the interactions of individual organisms, or individual species, with their environments and "synecology" to designate the study of groups of organisms treated as units. See Eugene P. Odum, *Fundamentals of Ecology*, 3rd ed. (Philadelphia: Saunders, 1971), p. 6. "Autecology" is often used synonymously with "physiological ecology," although the meaning of the latter, depending on the author, is sometimes more restrictive.
3. Hans Kniep, "Ernst Stahl," *Berichte der Deutschen Botanischen Gesellschaft* 37 (1919):(101). (Obituary notices in the *Berichte der Deutschen Botanischen Gesellschaft* always appear in a supplemental section at the end of each volume, with page numbers in parentheses to set them off from the normal pagination.)
4. Eugenius Warming, *Oecology of Plants: An Introduction to the Study of Plant-Communities*, trans. and ed. Percy Groom and I. B. Balfour (Oxford: Clarendon Press, 1909), p. 2. The original Danish textbook appeared in 1895.
5. Charles Elton, *Animal Ecology* (New York: Macmillan, 1927), p. 3.
6. See Stephen Jay Gould and Richard Lewontin, "The Spandrels of San Marco and the Panglossian Paradigm: A Critique of the Adaptationist Programme," *Proceedings of the Royal Society of London*, B, 205 (1979):581–98; but also Gould's more popular publications – for example, *Hen's Teeth and Horses' Toes: Further Reflections in Natural History* (New York: Norton, 1984), pp. 13–14, 155–66, and "The Mysteries of the Panda," *New York Review of Books* 32 (August 15, 1985):14–16.

1. Botany in Germany, 1850–1880: the making of a science and a profession

1. David E. Allen, for example, deplores the changes toward professionalization that took place in British biology at the end of the nineteenth century, attributing these new developments to the influence of the Germans. Allen writes: "many of the new generation of scientists became not merely dedicated professionals, but militant anti-amateurs; in the study of nature, choosing to lead a cellular existence of their own, bent firmly over their microscopes and locked in their laboratories."

The Naturalist in Britain: A Social History (London: Allen Lane–Penguin Books, 1976), p. 184.

2. One of the best sources for the history of botany in the nineteenth century, and German developments in particular, is still Julius Sachs, *History of Botany (1530–1860)*, trans. Henry E. F. Garnsey (Oxford: Clarendon Press, 1890), in conjunction with the sequel, J. Reynolds Green, *A History of Botany, 1860–1900* (Oxford: Clarendon Press, 1909). A. G. Morton, *History of Botanical Science: An Account of the Development of Botany from Ancient Times to the Present Day* (London: Academic Press, 1981), provides a very helpful overview of the whole history of botany, with Chs. 9 and 10 dealing in considerable detail with the nineteenth century. Karl Mägdefrau's *Geschichte der Botanik: Leben und Leistung grosser Forscher* (Stuttgart: Fischer, 1973) is a very readable and concise work emphasizing developments in Germany.
3. There are excellent discussions of the scientific contributions of Schleiden and Hofmeister in more detail than they will be presented here, in John Farley, *Gametes and Spores: Ideas About Sexual Reproduction, 1750–1914* (Baltimore: Johns Hopkins University Press, 1982), pp. 47–52, 82–100, and in Morton, *History of Botanical Science*, pp. 377–404.
4. Matthias Jacob Schleiden, *Gründzuge der wissenschaftlichen Botanik nebst einer methodologischen Einleitung* (Leipzig: Engelmann, 1842–3), 1:xi.
5. Ibid. 1:6.
6. For further discussion of Schleiden's philosophy of science see G. Buchdahl, "Leading Principles and Induction: The Methodology of Matthias Schleiden," in *Foundations of Scientific Method: The Nineteenth Century*, ed. R. N. Giere and R. S. Westfall (Bloomington: Indiana University Press, 1973), pp. 23–52, and the Introduction by Jacob Lorch in the recent reprint of Schleiden's textbook, *Principles of Scientific Botany or Botany as an Inductive Science*, trans. Edwin Lankester (1849; rpt. New York: Johnson Reprint Corp., 1969), pp. ix–xxxiv.
7. Karl F. Schimper, "Vorträge über die Möglichkeit eines Verstandnisses der Blattstellung," *Flora* 18 (1835):145–92; Alexander Braun, "Vergleichende Untersuchungen über die Ordnung der Schuppen an den Tannenzapfen, als Einleitung zur Untersuchung der Blattstellung überhaupt," *Nova acta Academiae Caeserae Leopoldino Carolinae Germanicae naturae curiosorum* 15 (1831):195–402; and Alexander von Humboldt, "Sur les lois que l'on observe dans la distribution des formes végétales," in *Dictionnaire des Sciences Naturelles* (Paris: Le Normant, 1820), 18:422–36.
8. Schleiden, *Principles of Scientific Botany*, p. 457.
9. The relationships among *Naturphilosophie*, the Kantian tradition in German science, and the new mechanistic biology are treated at length by Timothy Lenoir in *The Strategy of Life: Teleology and Mechanics in Nineteenth Century German Biology* (Dordrecht and Boston: D. Reidel, 1982). Schleiden later attacked the mechanistic materialists directly in a pamphlet, *Ueber den Materialismus der neuen deutschen Naturwissenschaft: Sein Wesen und seine Geschichte* (Leipzig: Engelmann, 1863).
10. Wilhelm Hofmeister, *Vergleichende Untersuchungen der Keimung, Entfaltung und Fruchtbildung höherer Kryptogamen (Moose, Farrn, Equisetaceen, Rhizocarpeen, und Lycopodiaceen) und der Samenbildung der Coniferen* (Leipzig: Engelmann, 1851), p. 1. Hofmeister wrote, but never published, a second edition of this work; but the manuscript was translated into English and published posthumously as *On the Germination, Development, and Fructification of the*

Higher Cryptogamia, and on the Fructification of the Coniferae, trans. F. Currey (London: John Murray, 1862).

11. Hofmeister, *Vergleichende Untersuchungen*, pp. 139–45. Modern botany textbooks refer to the two generations as "gametophyte" (haploid) and "sporophyte" (diploid), respectively.
12. Sachs, *History of Botany*, p. 200. The 1849 Hofmeister work to which Sachs refers is "Ueber die Fruchtbildung und Keimung der höheren Cryptogamen," *Botanische Zeitung* 7 (1849):793–800.
13. Carl Nägeli, "Wachstumsgeschichte von *Delassaria* und der Laub- und Lebermoose," in *Zeitschrift für wissenschaftliche Botanik*, ed. M. J. Schleiden and C. Nägeli (Zürich: Meyer and Zelle, 1845), 2:121–210.
14. Hugo von Mohl, *Gründzuge der Anatomie und Physiologie der vegetabilischen Zelle* (Braunschweig: F. Vieweg, 1851); *Principles of the Anatomy and Physiology of the Vegetable Cell*, trans. Arthur Henfry (London: John van Voorst, 1852).
15. Von Mohl, *Principles*, p. iv.
16. Schleiden, *Principles of Scientific Botany*, p. 575. The quotation is from the section "On the Use of the Microscope in Botanical Investigations," pp. 575–95. The translator, Edwin Lankester, left out most of Schleiden's methodological introduction in the English edition, but he included this section on microscopy, part of that introduction, in an appendix.
17. See S. Bradbury, *The Evolution of the Microscope* (Oxford: Pergamon Press, 1979), pp. 184–204; Arthur Hughes, "Studies in the History of Microscopy, I. The Influence of Achromatism," *Journal of the Royal Microscopical Society* 75 (1955):1–22; and Hughes, *A History of Cytology* (London: Abelard–Schuman, 1959), pp. 1–28. Recently, Brian Ford has argued that the quality of the image produced by the achromatic compound microscopes was no better, and possibly worse, than that produced by single-lens instruments at the time. See Brian J. Ford, *Single Lens: The Story of the Simple Microscope* (New York: Harper & Row, 1985), pp. 121–30.
18. Johannes Purkyně to R. Wagner, reprinted and trans. in Henry J. John, *Jan Evangelista Purkyně, Czech Scientist and Patriot, 1787–1869* (Philadelphia: American Philosophical Society, 1959), p. 28.
19. Hughes, *History of Cytology*, pp. 9–10; Everett Mendelsohn, "The Emergence of Science in Nineteenth-Century Europe," in *The Management of Science*, ed. Karl Hill (Boston: Beacon Press, 1964), pp. 18–21; but see especially William Coleman, "Prussian Pedagogy: Purkyně at Breslau, 1823–1839," in Coleman and F. L. Holmes, eds., *The Investigative Enterprise: Experimental Physiology in Nineteenth-Century Germany* (Berkeley: University of California Press, 1988), pp. 15–64.
20. Eduard Strasburger, "Botanik," in *Die Deutschen Universitäten*, ed. Wilhelm Lexis (Berlin: A. Asher, 1893), 2:89–90. For information regarding the dates of the establishment of botanical institutes at the various universities and for statistics concerning funding, see pp. 161–78 in this volume, as well as Lexis, ed., *Das Unterrichtswesen im Deutschen Reich, I. Die Deutschen Universitäten* (Berlin: A. Asher, 1904).
21. Charles E. McClelland, *State, Society, and University in Germany, 1700–1914* (Cambridge: Cambridge University Press, 1980), pp. 204–5, 280–7. As an example of this shift of emphasis toward research, McClelland points out that the largest single item in the budget at the University of Berlin in 1860 was faculty salaries, whereas by 1870 the largest budget item was institutes and seminars.

22. See E. G. Pringsheim, *Julius Sachs: der Begründer der neueren Pflanzenphysiologie, 1832–1897* (Jena: G. Fischer, 1932), pp. 1–30.
23. Hughes, *History of Cytology*, pp. 12–17. There is a thorough discussion of the improvements in staining and sectioning techniques in Brian Bracegirdle, *A History of the Microtome and the Development of Tissue Preparation* (London: Heinemann, 1978).
24. The most well known of these were his auxanometer, for measuring very tiny increments of growth, and his klinostat, for eliminating the effects of gravitation on growing plants. These and many other such devices are described in Pringsheim, *Julius Sachs*, pp. 218–30.
25. There is very little German literature devoted directly to a discussion of the physical nature and organizational structure of the institutes. The best descriptions are to be found in memoirs of botanists, such as Gottlieb Haberlandt, *Erinnerungen: Bekenntnisse und Betrachtungen* (Berlin: Springer, 1933), and Alexander Tschirch, *Erlebtes und Erstrebtes: Lebenserinnerungen* (Bonn: Friedrich Cohn, 1921). Fortunately, three British botanists who visited German laboratories in the 1870s and 1880s were impressed sufficiently to describe these facilities in some detail later: S. H. Vines, "Reminiscences of German Botanical Laboratories in the 'Seventies and 'Eighties of the Last Century"; D. H. Scott, "German Reminiscences of the Early 'Eighties"; and F. O. Bower, "English and German Botany in the Middle and Towards the End of the Last Century," *The New Phytologist* 24 (1925):1–8, 9–16, 129–37.
26. By contrast, in England, where universities made little provision for laboratory study until the 1870s and 1880s, taxonomy remained the central concern of botanists until fairly late in the nineteenth century. There was perhaps an additional economic factor: Earlier in the century, the best English microscopes were quite expensive, and therefore were available only to the amateur with sufficient means (Bradbury, *Evolution of the Microscope*, p. 204). However, the price of microscopes may simply have reflected English interests and tastes. In any event, a host of socioeconomic factors probably served to limit the market for cheap, accurate microscopes in England. Not the least of these, insofar as botany was concerned, was the preoccupation of British botanists with the systematic work resulting from the early acquisition of colonial territories. For a discussion of the relationship between colonial expansion and botany in Britain, see Lucille H. Brockway, *Science and Expansion: The Role of the British Royal Botanic Gardens* (New York: Academic Press, 1979).
27. Some German botanists even credited Hofmeister with anticipating Darwin's theory of descent. For example, Sachs stated in his *History of Botany*, p. 202: "That which Haeckel, after the appearance of Darwin's book, called the phylogenetic method, Hofmeister had long before actually carried out, and with magnificent success." However, in his biography of Hofmeister, Karl Goebel took pains to point out that Hofmeister had no intention of interpreting his work phylogenetically. Goebel, *Wilhelm Hofmeister: The Work and Life of a Nineteenth Century Botanist*, trans. H. M. Bower (London: Ray Society, 1926), pp. 60–2.
28. The second edition of Schleiden's textbook, as well as the third and fourth, carried the additional title *Die Botanik als inductive Wissenschaft*, 2nd ed. (Leipzig: Engelmann, 1845–6), 3rd ed. (1849), 4th ed., a reprint of the 3rd (1861).
29. Pringsheim, *Untersuchungen über den Bau und die Bildung der Pflanzenzelle, Erste Abteilung: Grundlinien einer Theorie der Pflanzenzelle* (Berlin: Hirschwald, 1854).

30. Hofmeister, *Die Lehre von der Pflanzenzelle* (Leipzig: Engelmann, 1867), and *Allgemeine Morphologie der Gewächse* (1868); Anton de Bary, *Morphologie und Physiologie der Pilze, Flechten, und Myxomyceten* (1866); Julius Sachs, *Handbuch der Experimental-Physiologie der Pflanzen: Untersuchungen über die allgemeinsten Lebensbedingungen der Pflanzen: und die Functionen ihrer Organe* (1865).
31. Julius Sachs, *Lehrbuch der Botanik nach dem gegenwärtigen Stand der Wissenschaft* (Leipzig: Engelmann, 1868; 4th ed., 1874). The English translation is based on the third, and part of the fourth, editions: *Text-Book of Botany, Morphological and Physiological,* trans. A. W. Bennet and W. T. Thiselton Dyer (Oxford: Clarendon Press, 1875).
32. Ernst G. Pringsheim, *Julius Sachs: Der Begründer der neueren Pflanzenphysiologie, 1832–1897* (Jena: G. Fischer, 1932), pp. 1–30. See also Karl Goebel, "Julius Sachs," *Science Progress,* 7 (1898):150–73, reprinted in *Science* 7 (1898):662–8, 695–702; and Fritz Noll, "Julius von Sachs," *Botanical Gazette* 25 (1898):1–12.
33. Bower, "English and German Botany," p. 132.
34. Charles Bessey, *Botany for High Schools and Colleges* (New York: Holt, 1880); Charles E. Ford, "Botany Texts: A Survey of Their Development in American Higher Education, 1643–1906," *History of Education Quarterly* 4 (1964):59–71.
35. Goebel, "Julius Sachs," pp. 162–3.
36. See Philip J. Pauly, *Controlling Life: Jacques Loeb and the Engineering Ideal in Biology* (New York: Oxford University Press, 1987), pp. 34–6.
37. British botanists Sidney Vines said of Sachs: "His only rival as a lecturer, in my experience, has been Professor Huxley." D. H. Scott, Vines's traveling companion in Germany, said simply: "Sachs was the best lecturer I ever heard." Vines, "Reminiscences of German Botanical Laboratories," p. 4; Scott, "German Reminiscences," p. 11.
38. See Pringsheim, *Julius Sachs;* also Karl Goebel, "Ueber Leben und Werk von Julius Sachs," *Flora* 84 (1897):101–30. Sachs later produced a more comprehensive physiology textbook, *Vorlesungen über Pflanzenphysiologie* (Leipzig: Engelmann, 1883; 2nd ed., 1887), trans. Marshall Ward, *Lectures on the Physiology of Plants* (Oxford: Clarendon Press, 1887). This textbook was eventually superseded by that of Sachs's student Wilhelm Pfeffer, *Pflanzenphysiologie: Ein Handbuch der Lehre vom Stoffwechsel und Kraftwechsel der Pflanzen* (Leipzig: Engelmann, 1881; 2nd ed., 1897–1904), trans. Alfred J. Ewart, *The Physiology of Plants: A Treatise upon the Metabolism and Sources of Energy in Plants,* 3 vols. (Oxford: Clarendon Press, 1899–1905).
39. Nägeli, *Entstehung und Begriff der Naturhistorischen Art* (Munich: Königl. Akademie, 1865); Strasburger, "Ueber die Bedeutung Phylogenetischer Methoden für die Erforschung lebender Wesen," *Jenaische Zeitschrift für Wissenschaft* 8 (1874):56–80.
40. Sachs, *Lehrbuch der Botanik* (1874), pp. 894–920.
41. Heinz Haushofer, *Die deutsche Landwirtschaft im technischen Zeitalter,* Band 5, *Deutsche Agrargeschichte,* ed. G. Franz (Stuttgart: Verlag Eugen Ulmer, 1963), p. 163; H. Thiel, *Die deutsche Landwirtschaft auf der Weltausstellung in Paris 1900* (Bonn: Universitäts-Buchdruckerei von Carl Georgi, 1900), pp. 92–3; Friedrich-Wilhelm Henning, *Landwirtschaft und ländliche Gesellschaft in Deutschland,* vol. 2 (Paderborn: Ferdinand Schöningh, 1978), p. 89.
42. Sachs, *Ueber den gegenwärtigen Zustand der Botanik in Deutschland* (Würzburg: F. E. Thein, 1872), p. 18.

43. Lexis, *A General View of the History and Organization of Public Education in the German Empire*, trans. G. J. Tamson (Berlin: A. Asher, 1904), p. 50. For statistics on botany professorships in the late nineteenth century, see Lexis, *Die Deutschen Universitäten*, 2:171.
44. Fritz Ringer, *The Decline of the German Mandarins: The German Academic Community, 1890–1933* (Cambridge: Harvard University Press, 1969), pp. 25–42; also Ringer, "The German Academic Community," in *The Organization of Knowledge in Modern America*, ed. A. Oleson and J. Voss (Baltimore: Johns Hopkins University Press, 1979), pp. 409–29; and McClelland, *State, Society, and University in Germany*, Chs. 6 and 7.
45. Lexis, *A General View*, p. 34.
46. Lexis, *A General View*, p. 14; Lexis, *Die Deutschen Universitäten*, 2:171; Lexis, *Das Unterrichtswesen im Deutschen Reich*, 1:325. Ringer, "The German Academic Community," p. 419, indicates that in the period from 1864 to 1910 the percentage of full professors in the sciences at German universities declined overall from fifty-two to thirty-five.
47. Ringer, *Decline of the German Mandarins*, pp. 37–8; Lexis, *Das Unterrichtswesen*, 1:42–7. The income figures apply to the 1890s. For a discussion of German university teaching and the system of promotions, see Friedrich Paulsen, *The German Universities and University Study*, trans. F. Thilly and W. W. Elwang (New York: Charles Scribner's Sons, 1906), esp. pp. 163–71, and Max Weber, "Science as a Vocation," trans. H. H. Gerth and C. Wright Mills in *Max Weber: Essays in Sociology* (Oxford: Oxford University Press, 1946), p. 129–56.
48. Sachs, *Ueber den gegenwärtigen Zustand der Botanik*, pp. 6–7.
49. Strasburger, "Botanik," p. 73.
50. Asa Gray, review of *Botany for High Schools and Colleges*, by C. E. Bessey, *American Journal of Science* 120 (1880):337.
51. See Richard Overfield, "Charles E. Bessey: The Impact of the 'New' Botany on American Agriculture, 1880–1910," *Technology and Culture* 16 (1975):162–81; W. G. Farlow, "The Change from the Old to the New Botany in the United States," *Science* 37 (1913):79–86; and the Bower, Scott, and Vines reminiscences (note 25, this chapter).
52. Strasburger, "Botanik," p. 73.
53. Sachs, *Ueber den gegenwärtigen Zustand der Botanik*, p. 7.

2. Schwendener and Haberlandt: the birth of physiological plant anatomy

1. Biographical information on Schwendener was drawn mainly from Albrecht Zimmermann, "Simon Schwendener," *Berichte der Deutschen Botanischen Gesellschaft* 40 (1922):(55)–(76), which includes a brief autobiographical sketch. Additional sources: Gottlieb Haberlandt, "Gedächtnisrede auf Simon Schwendener," *Abhandlungen der Preussischen Akademie der Wissenschaften zu Berlin* (1919):3–12; idem, "Gedächtnisrede," *Berichte der Deutschen Botanischen Gesellschaft* 47 (1929):3–19; Alexander Tschirch, *Erlebtes und Erstrebtes, Lebenserinnerungen* (Bonn: Friedrich Conn, 1921), pp. 168–73.
2. Simon Schwendener, *Ueber die periodischen Erscheinungen der Natur, insbesondere der Pflanzenwelt* (Zürich: E. Kiesling, 1856).
3. Carl Nägeli and Simon Schwendener, *Das Mikroskop*, 2 parts (Leipzig: Engelmann, 1865–7). A single-volume edition was published by Engelmann as *Das Mikroskop: Theorie und Anwendung desselben* in 1877.

4. Schwendener's work on lichens appeared in several journal articles throughout the 1860s, beginning with "Untersuchungen über den Flechtenthallus," *Beiträge zum wissenschaftlichen Botanik* 2 (1860):109–86. The Zimmermann obituary in *Berichte der Deutschen Botanischen Gesellschaft,* as is customary for that journal, includes a complete bibliography of Schwendener's published works.
5. Schwendener, *Das mechanische Princip in anatomischen Bau der Monocotylen, mit vergleichenden Ausblicken auf die übrigen Pflanzenklassen* (Leipzig: Engelmann, 1874).
6. Schwendener, autobiographical sketch in Zimmermann, "Simon Schwendener," p. (58). The reviews by G[regor] K[rause], *Botanische Zeitung* 32 (1874):746, and E. Loew, *Botanischer Jahresbericht* 2 (1874):445–50, treated Schwendener's work favorably, if not enthusiastically.
7. Anton de Bary, *Comparative Anatomy of the Vegetative Organs of the Phanerogams and Ferns,* trans. F. O. Bower and D. H. Scott (Oxford: Clarendon Press, 1884), p. 416; idem, *Vergleichende Anatomie der Vegetationsorgane der Phanerogamen und Farne* (Leipzig: Engelmann, 1877).
8. Sachs, "Bemerkungen zum Anpassungen," in E. G. Pringsheim, *Julius Sachs, der Begründer der neueren Pflanzenphysiologie, 1832–1897* (Jena: G. Fischer, 1932), pp. 153–4. Pringsheim discovered this paper among Sachs's notes and published it posthumously in his biography of Sachs.
9. Schwendener, autobiographical sketch in Zimmermann, "Simon Schwendener," p. (58).
10. Herman von Guttenberg, "Gottlieb Haberlandt," *Phyton* 6 (1955):1–14; Haberlandt, *Erinnerungen: Bekenntnisse und Betrachtungen* (Berlin: Springer, 1933); and Otto Renner, obituary notice in *Jahrbuch der bayerischen Akademie der Wissenschaften* (1944), pp. 258–61.
11. Haberlandt, *Erinnerungen,* pp. 57–8.
12. Haberlandt, "Untersuchungen über die Winterfärbung ausdauernder Blatter," *Sitzungsberichte der Akademie der Wissenschaften zu Wien,* math.-naturw. Classe, 73 (1876), 1:267–96. The Guttenberg biographical sketch includes a complete bibliography of Haberlandt's work.
13. Haberlandt, *Die Schutzeinrichtungen in der Entwickelung der Keimpflanze: Eine biologische Studie* (Vienna: C. Gerold's Sohn, 1877).
14. Haberlandt, *Erinnerungen,* p. 65.
15. Haberlandt, *Erinnerungen,* p. 187. Kerner, *Das Planzenleben der Donauländer* (Innsbruck: Wagner, 1863).
16. Haberlandt, *Erinnerungen,* p. 77.
17. S. H. Vines, "Reminiscences of German Botanical Laboratories," and F. O. Bower, "English and German Botany in the Middle and Towards the End of the Last Century," *The New Phytologist* 24 (1925):1–8, 129–37.
18. Haberlandt, *Erinnerungen,* pp. 65–6.
19. Reported by Haberlandt, ibid., p. 73.
20. Haberlandt, *Die Entwickelungsgeschichte des mechanischen Gewebesystems der Pflanzen* (Leipzig: Engelmann, 1879).
21. In 1945, nearly ninety years old, Haberlandt fled Berlin with his wife during the worst of the bombing to stay with friends in the country. They returned just before the end of the war, and the two were separated when they attempted to find solace in a house of charity. Considered beyond help, the exhausted Haberlandt was cast out of the house and soon died. Guttenberg, "Haberlandt," p. 6.
22. The only clearly critical treatment of Schwendener's *Mechanische Princip* that I have been able to find appeared in a book by Paul Falkenberg, a young Ph.D. from Göttingen, whose 1875 dissertation concerned the structure of the vegetative

organs of monocotyledons. Falkenberg's *Vergleichende Untersuchungen über den Bau der Vegetationsorgane der Monocotyledonen* (Stuttgart: F. Enke, 1876) contains specific critical references to Schwendener's book, somewhat along the lines of Sachs's criticism (see p. 30, this chapter). No one, including Schwendener, took this criticism as a serious threat. Haberlandt mentioned Falkenberg briefly in *Entwickelungsgeschichte*, but he characterized his specific criticisms as polemical and declined to address them.

23. Haberlandt, *Entwickelungsgeschichte*, p. 69.
24. Review of *Entwickelungsgeschichte des mechanischen Gewebesystems der Pflanzen* in *Botanische Zeitung* 37 (1879):335–40.
25. Haberlandt, *Entwickelungsgeschichte*, p. 1.
26. Ibid., p. 2.
27. Review of *Entwickelungsgeschichte*, p. 337.
28. Ibid., p. 340.
29. Simon Schwendener, "Antrittsrede," Inaugural Address to the Berlin Academy of Sciences, July 8, 1880, *Monatsberichte der Preussischen Akademie der Wissenschaften zu Berlin* (1880), pp. 621–3.
30. Haberlandt, "Vergleichende Anatomie des assimilatorischen Gewebesystems der Pflanzen," *Jahrbücher für wissenschaftliche Botanik* 13 (1881):74–188; idem, "Die physiologischen Leistungen der Pflanzengewebe," in *Handbuch der Botanik*, ed. August Schenk (Breslau: E. Trewendt, 1882), pp. 558–693; idem, *Physiologische Pflanzenanatomie im Grundriss dargestellt* (Leipzig: Engelmann, 1884).
31. Haberlandt, "Vergleichende Anatomie," p. 82.
32. In *Physiologische Pflanzenanatomie* Haberlandt first discusses these principles in general terms (pp. 23–5), and he later reintroduces them in the sections dealing with particular plant systems.
33. *Physiologische Pflanzenanatomie* appeared in six German editions between 1884 and 1924. The English translation, *Physiological Plant Anatomy*, trans. M. Drummond (London: Macmillan, 1914), is based on the fourth (1909) edition.
34. This three part classification scheme appeared in the first edition of Sachs, *Lehrbuch der Botanik*, and Sachs repeated it in all subsequent editions. The most readily available, the fourth edition (1874), includes an eighty-page section on tissue classification (pp. 70–151).
35. de Bary, *Comparative Anatomy*, pp. 116–18, 406–12.
36. Haberlandt, *Physiologische Pflanzenanatomie*, pp. 180–91. This discussion is relatively unchanged in the English edition, *Physiological Plant Anatomy*, pp. 276–91.
37. Haberlandt, *Erinnerungen*, p. 96.
38. Haberlandt, *Physiological Plant Anatomy* (1914), p. 12.

3. Overtures to Darwinism

1. See E. S. Russell, *Form and Function: A Contribution to the History of Animal Morphology* (London: John Murray, 1916; rpt. Chicago: University of Chicago Press, 1982), pp. 45–51; also the Introduction to the 1982 edition by George V. Lauder. Agnes Arber has translated Goethe's major contribution to plant morphology, *Versuch die Metamorphose der Pflanzen zu erklären*, in *Goethe's Botany* (Waltham, MA: Chronica Botanica, 1946), pp. 91–115.
2. Gottlieb Haberlandt, *Erinnerungen: Bekenntnisse und Betrachtungen* (Berlin: Springer, 1933), p. 97.

3. Haberlandt, "Zur Geschichte der physiologischen Pflanzenanatomie," *Berichte der Deutschen Botanischen Gesellschaft* 40 (1922):157.
4. Otto Warburg, review of *Physiologische Pflanzenanatomie* in *Botanische Zeitung* 45 (1885):26–32. This is the botanist Otto Warburg (1859–1958), not to be confused with the better-known biochemist by the same name.
5. Most German botanical journals offered descriptive rather than critical reviews of current literature. *Botanische Zeitung* was the important exception, and Otto Warburg's somewhat restrained comments constituted the harshest public criticism Haberlandt's book was to receive.
6. Haberlandt, *Erinnerungen,* pp. 97–8.
7. Haberlandt, *Physiologische Pflanzenanatomie im Grundriss dargestellt* (Leipzig: Engelmann, 1884), p. 173.
8. Ibid., pp. 17–18. Parts of this quotation appeared verbatim in "Vergleichende Anatomie des assimilatorischen Gewebesystems der Pflanzen," *Jahrbücher für wissenschaftliche Botanik* 13 (1881):83.
9. Haberlandt, "Vergleichende Anatomie," p. 83.
10. Haberlandt, *Erinnerungen,* p. 101.
11. In this section I am making liberal use of the unpublished dissertation of William Montgomery, "Evolution and Darwinism in German Biology, 1800–1883" (Ph.D. diss., University of Texas, 1975). Montgomery summarized the main points of this work in "Germany," in *The Comparative Reception of Darwinism,* ed. Thomas F. Glick (Austin: University of Texas Press, 1972), pp. 81–116. I have also relied upon the treatment of the immediate reception of Darwinian biology provided by Pierce C. Mullen in "The Preconditions and Reception of Darwinian Biology in Germany, 1800–1870" (Ph.D. diss., University of California, 1964), pp. 268–85.
12. Karl Ernst von Baer, "The Controversy over Darwinism" [1876], trans. David L. Hull, in Hull, *Darwin and His Critics: The Reception of Darwin's Theory of Evolution by the Scientific Community* (Chicago: University of Chicago Press, 1973), p. 421.
13. There is a discussion of this incident and its ramifications in Alfred Kelly, *The Descent of Darwin: The Popularization of Darwinism in Germany, 1860–1914* (Chapel Hill: University of North Carolina Press, 1981), pp. 57–74.
14. Montgomery, "Evolution and Darwinism in German Biology," pp. 147–53.
15. Montgomery, "Evolution and Darwinism in German Biology," pp. 151–3; Kelly, *The Descent of Darwin,* pp. 18–23; Paul J. Weindling, "Darwinism in Germany," in David Kohn, ed., *the Darwinian Heritage* (Princeton: Princeton University Press, 1985), pp. 685–98.
16. Haeckel, *Generelle Morphologie der Organismen: allgemeine Grundzüge der organischen Formen-Wissenschaft mechanisch begründet durch die von Charles Darwin reformirte Descendenz-Theorie,* 2 vols. (Berlin: Reimer, 1866); *Natürliche Schopfungsgeschichte* (Berlin: Reimer, 1868). See Montgomery, "Evolution and Darwinism," pp. 201–2, and Weindling, "Darwinism in Germany," pp. 692–6.
17. Darwin, *Ueber die Entstehung der Arten im Thier-und Pflanzenreich durch natürliche Züchtung, oder Enthaltung der vervollkommneten Rassen im Kampfe um's Dasein,* trans. H. G. Bronn (Stuttgart: Schweizerbart, 1860). Bronn's remarks, entitled "Schlusswort der Übersetzers," follow the text.
18. Darwin, *Ueber die Entstehung der Arten durch natürliche Züchtwahl; oder, die Erhaltung der begünstigsten Rassen im Kampfe um's Dasein,* trans. J. V. Carus (Stuttgart: Schweizerbart, 1867); see Kelly, *Descent of Darwin,* pp. 20–1, and Montgomery, "Germany," pp. 91–3.

19. Carl Nägeli, *Entstehung und Begriff der historisches Art* (Munich: Königl. Akademie, 1865); idem, "Ueber den Einfluss der äusseren verhältnisse auf die Varietätenbildung im Pflanzenreiche," *Sitzungsberichte der Akademie der Wissenschaften zu München* 2 (1865):228–32, 258–60, 277–84. This second work is actually a refutation of the role of external influences on species formation in the sense of Lamarck, but Nägeli believed that no external influence, whether through direct action or natural selection, could account for the production of new species.
20. See Peter J. Bowler, "Theordor Eimer and Orthogenesis: Evolution by 'Definitely Directed Variation.'" *Journal of the History of Medicine* 34 (1979):40–73, and the chapter on orthogenesis in Bowler, *The Eclipse of Darwinism: Anti-Darwinian Evolution Theories in the Decades around 1900* (Baltimore: Johns Hopkins University Press, 1983), pp. 141–81.
21. William Coleman, "Morphology Between Type Concept and Descent Theory," *Journal of the History of Medicine* 31 (1976):149–75.
22. William Coleman, "Cell, Nucleus, and Inheritance: An Historical Study," *Proceedings of the American Philosophical Society* 109 (1965):124–58.
23. Haberlandt, "Gedächtnisrede auf Simon Schwendener," *Abhandlungen der Preussichen Akademie der Wissenschaften zu Berlin* (1919):10–11.
24. Haberlandt, *Erinnerungen*, p. 42.

4. Schwendener's circle: botanical "comrades-in-arms"

1. Constanz Gangenmöller, "Wilhelm Hofmeister: Biographical Supplement," in Karl Goebel, *Wilhelm Hofmeister: The Work and Life of a Nineteenth Century Botanist*, trans. H. M. Bower (London: Ray Society, 1926), pp. 196–202.
2. Gottlieb Haberlandt, *Erinnerungen: Bekenntnisse und Betrachtungen* (Berlin: Springer, 1933), p. 68.
3. Ibid., p. 74.
4. Wilhelm Lexis, *A General View of the History and Organization of Public Education in the German Empire*, pp. 14, 39; *Die Deutschen Universitäten*, ed. Lexis (Berlin: A. Asher, 1893), 2:175.
5. See Simon Schwendener, "Festrede," *Berichte der Deutschen Botanischen Gesellschaft* 25 (1907):(21)–(31); and Heinz Degen, "Die Entstehung der Deutschen Botanischen Gesellschaft," *Naturwissenschaftliche Rundschau* 27 (1974):333–40.
6. Haberlandt, *Erinnerungen*, p. 74.
7. Zimmermann, "Simon Schwendener," *Berichte der Deutschen Botanischen Gesellschaft* (1922):(68).
8. Alexander Tschirch, *Erlebtes und Erstrebtes: Lebenserinnerungen* (Bonn: Friedrich Cohn, 1921), p. 71.
9. Ibid., p. 169.
10. Otto Reinhardt, "Georg Volkens," *Berichte der Deutschen Botanischen Gesellschaft* 35 (1917):(66).
11. Albrecht Zimmermann, "Ueber mechanische Einrichtungen zur Verbreitung der Samen und Früchte mit besonderer Berücksichtigung der Torsionserscheinungen," *Jahrbücher für wissenschaftliche Botanik* 12 (1881):542–77; Hermann Ambronn, "Zur mechanik des Windens," *Berichte der sächsischen Gesellschaft der Wissenschaft zu Leipzig* 36 (1884):136–84; Gustav Krabbe, "Ueber das Wachsthum des Verdickungsringes und der jungen Holzzellen in seiner Abhängigkeit von Druckwirkungen," *Abhandlungen der Akademie der Wissenschaften zu Berlin* (1882), pp. 1–83.

12. Max Westermaier, "Untersuchung über den Bau und die Function des pflanzlichen Hautgewebes," *Sitzungsberichte der Akademie der Wissenschaften zu Berlin* (1882), pp. 837–43; idem, "Zur physiologischen Bedeutung des Gerbstoffes in den Pflanzen," ibid. (1885), pp. 1115–26; Nordal Wille, "Bidrag til Algernes physiologiske Anatomi," "Svenska Vetenskaps-Akademiens Handlingar 21 (1886):1–78.
13. Emil Heinricher, "Ueber isolateralen Blattbau mit besonderer Berücksichtigung der europäischen, speziell der deutschen Flora: Ein Beitrag zur Anatomie und Physiologie der Laubblatter," *Jahrbücher für wissenschaftliche Botanik* 15 (1884):502–67.
14. Henry Potonié was August Eichler's assistant at the Berlin Botanical Garden from 1880 to 1883. Although he spent considerable time at Schwendener's institute in the 1880s, his own interests turned to paleobotany. He taught paleobotany at the Royal School of Mines in Berlin during the 1890s, and in 1900 he was appointed Germany's state geologist. Potonié edited the *Naturwissenschaftliche Wochenschrift* from its inception in 1888 until his death in 1913, always devoting ample space in its pages to discussions of the work of Schwendener and his students. See W. Gothan, "H. Potonié, *Berichte der Deutschen Botanischen Gesellschaft* 31 (1913):(127)–(136).
15. C. Correns and H. Morstatt, "Albrecht Zimmermann," *Berichte der Deutschen Botanischen Gesellschaft* 49 (1931):(220)–(243).
16. Adolf Sperlich, "Emil Heinricher," *Berichte der Deutschen Botanischen Gesellschaft* 52 (1934):(188)–(205). Although Wagner was a strong Lamarckian regarding the mechanism of evolution, the influence of physiological plant anatomy is nevertheless quite evident in his study of alpine plants, "Zur Kenntnis des Blattbaues der Alpenpflanzen und dessen biologischer Bedeutung," *Sitzungsberichte der Akademie der Wissenschaften zu Wien,* math.-naturw. Classe, 101 (1892), 1:487–548.
17. Alexander von Humboldt and Aimé Bonpland, *Essai sur la géographie des plantes* (Paris: F. Schoell, 1807; rpt. New York: Arno Press, 1977); Humboldt, "Sur les lois que l'on observe dans la distribution des formes végétales," in *Dictionnaire des sciences naturelles* (Paris: Le Normant, 1820), 18:422–36. For a discussion of Humboldt's contributions to plant geography, along with an account of the entire movement toward identifying natural plant provinces, see Janet Browne, *The Secular Ark: Studies in the History of Biogeography* (New Haven: Yale University Press, 1983), pp. 32–57.
18. August Grisebach, *Die Vegetation der Erde nach ihrer klimatischen Anordnung: ein Abriss der vergleichenden Geographie der Pflanzen,* 2 vols. (Leipzig: Engelmann, 1872). Grisebach's son collected his shorter works on plant geography in *Gesammelte Abhandlungen und kleinere Schriften zur Pflanzengeographie,* ed. Eduard Grisebach (Leipzig: Engelmann, 1880). The annual *Botanischer Jahresbericht,* also known as *Just's Botanischer Jahresbericht,* after its editor, Leopold Just, was published in Berlin from 1873 to 1913. For that time period it is perhaps the best source available for quick reference to published material on any given botanical topic in a given year. The most complete discussion of the development of the various branches of plant geography in the nineteenth century is to be found in Adolf Engler, "Die Entwickelung der Pflanzengeographie in den letzten hundert Jahren und weitere Aufgabe derselben," *Beiträge zum Gedächtniss der hundertjährigen Wiederkehr des Antritts von Alexander von Humboldt's Reise nach Amerika* Berlin: W. H. Köhl, 1899), pp. 1–247.
19. Biographical information on Tschirch was taken from T. Sabalitschka, "Alexander Tschirch," *Berichte der Deutschen Botanischen Gesellschaft* 59 (1941–2):(67)–(108), and from Tschirch's memoirs, *Erlebtes und Erstrebtes.*

20. Tschirch, *Erlebtes und Erstrebtes,* p. 170.
21. Tschirch, "Ueber einige Beziehungen des anatomischen Baues der Assimilationsorgane zu Klima und Standort, mit specieller Berücksichtigung des Spaltöffnungsapparates," *Linnaea* 43 (1881):139–252. In 1882 *Linnaea* was taken over by the University of Berlin and became the *Jahrbuch des botanischen Gartens zu Berlin.*
22. Tschirch, *Erlebtes und Erstrebtes,* p. 176.
23. Ibid., p. 176. Tschirch did not indicate whether he asked for the picture.
24. Tschirch, "Die Bedeutung der Blätter im Haushalte der Natur" [1893], reprinted in *Vorträge und Reden von A. Tschirch, gesammelt und herausgegeben von Schüler und Freunden* (Leipzig: Borntraeger, 1915), p. 55.
25. Tschirch, "Der javanische Urwald," *Jahrbuch der Geographischen Gesellschaft zu Bern* (1890), reprinted in Tschirch, *Vorträge und Reden,* pp. 222–35; idem, "Physiologische Studien über die Samen, insbesondere die Sangorgane derselben," *Annales du jardin botanique de Buitenzorg* 9 (1891):143–83.
26. Reinhardt, "Georg Volkens"; H. Harms, "Georg Volkens," *Verhandlungen des Botanischen Vereins der Provinz Brandenburg* 59 (1917):1–23. The latter work includes a twelve-page autobiographical sketch by Volkens annotated by Harms.
27. Georg Volkens, "Ueber Wasserauscheidung in liquider Form an den Blättern höherer Pflanzen," *Berlin Universität, Botanisches Garten, Jahrbuch* 2 (1883): 166–209.
28. Volkens, "Zur Kenntnis der Beziehungen zwischen Standort und anatomischen Bau der Vegetationsorgane," *Berlin Universität, Botanisches Garten, Jahrbuch* 3 (1884):1–46; Paul Sorauer, "Studien über Verdunstung," *Forschungen auf der Gebiet dem Agriculturphysik* 3 (1880):351–490.

5. Physiological anatomy beyond the Reich

1. For further discussion of the connection between the Berlin Academy of Sciences and imperialism, see Werner Hartkopf, *Die Akademie der Wissenschaften der DDR: Ein Beitrag zur ihrer Geschichte* (Berlin: Akademie-Verlag, 1975), pp. 55–8; and Conrad Grau, *Die Berliner Akademie der Wissenschaften in der Zeit des Imperialismus,* Part 1 (Berlin: Akademie-Verlag, 1975), Ch. 1.
2. Georg Volkens, autobiographical sketch in H. Harms, "Georg Volkens," *Verhandlungen des Botanischen Vereins der Provinz Brandenburg* 59 (1917):2.
3. Ibid.
4. Volkens, "Zur Flora der ägyptisch-arabischen Wüste: Eine Vorläufige Skizze," *Sitzungsberichte der Akademie der Wissenschaften zu Berlin,* phys.-math. Klasse 28 (1886):63–4.
5. Volkens, *Die Flora der ägyptisch-arabischen Wüste, auf Grundlage anatomisch-physiologischer Forschungen dargestellt* (Berlin: Gebrüder Borntraeger, 1887), p. 2.
6. Ibid., pp. 2–4.
7. Friedrich Kohl, *Die Transpiration der Pflanzen und ihre Einwirkung auf die Ausbildung pflanzlicher Gewebe* (Braunschweig: H. Bruhn, 1886); Hubert Leitgeb, "Beiträge zur Physiologie der Spaltöffnungsapparate," in *Mittheilungen aus dem botanischen institut zu Graz* (Jena: G. Fischer, 1886), 1:125–86; Gottlieb Haberlandt, *Physiologische Pflanzenanatomie im Grundriss dargestellt* (Leipzig: Engelmann, 1884); idem, "Ueber das Assimilationssytem," *Berichte der Deutschen Botanischen Gesellschaft* 4 (1886):(206)–(36); Ernst Stahl, "Ueber den Einfluss des sonnigen oder schattigen Standortes auf die Ausbildung der Laubblätter," *Jenaische Zeitschrift für Naturwissenschaft* 16 (1883):162–200.

8. Volkens, *Die Flora*, pp. 35–8.
9. Ibid., p. 39.
10. Ibid., pp. 75–6.
11. Ibid., pp. 34–5n.
12. A. Fischer, review of *Die Flora der ägyptisch-arabischen Wüste*, by G. Volkens, *Botanische Zeitung* 46 (1888):75–6.
13. Volkens, autobiographical sketch in Harms, "Volkens," p. 2.
14. Volkens, *Der Kilimandscharo: Darstellung der allgemeineren Ergebnisse eines fünfzehnmonatigen Aufenthalts im Dschaggalande* (Berlin: Geographische Verlagshandlung Dietrich Reiner, 1897).
15. Volkens, autobiographical sketch in Harms, "Georg Volkens," p. 7; see also Otto Reinhardt, "Georg Volkens," *Berichte der Deutschen Botanischen Gesellschaft* 35 (1917):(71)–(76).
16. Volkens published a short report on colonial agriculture in *Deutsches Kolonialblatt* 4 (1893):435–6 and thereafter in many subsequent issues. He was not so much an expert on colonial agriculture as a promoter; his publications were intended more to interest his countrymen in the plant resources of the colonies than to instruct them. See Volkens, "Die Botanische Zentralstelle für die Kolonien, ihre Zwecke und Ziele," *Jahresbericht der Vereinigung für angewandte Botanik* 5 (1907):32–48; idem, "Die Entwickelung des auf wissenschaftlicher Grundlage ruhenden landwirtschaftlichen Versuchswesens in den Kolonien," *Verhandlungen 3e Deutsch-Kolonialkongress* (1910), pp. 60–70; Volkens and Engler, "Die land-und forstwirtschaftlichen Versuchsstationen der deutschen Kolonien," *Congrès international d'agronomie coloniale et tropicale de Bruxelles* (Brussels: J. Goemaere, 1910).
17. Bernhard Zepernick and Else-Marie Karlsson, *Berlins Botanischer Garten* (Berlin: Haude and Spener, 1979), pp. 90–103. Reports on the activity of the Botanische Zentralstelle appeared regularly in the *Notizblatt des Königlichen botanischen Gartens und Museums zu Berlin* and in the *Denkschrift über die Entwicklung der deutschen Schutzgebiete in Afrika und der Südsee.*
18. Volkens, "Die Vegetation der Karolinen; mit besonderer Berücksichtigung der von Yap," *Botanische Jahrbücher* 31 (1901):412–77.
19. Volkens, *Laubfall und Lauberneuerung in den Tropen* (Berlin: Borntraeger, 1912).
20. For general background on the German colonial venture and its relationship to that of other European nations, see Raymond F. Betts, *The False Dawn: European Imperialism in the Nineteenth Century,* Ernst S. Dodge, *Islands and Empires: Western Impact on the Pacific and East Asia,* and Henry S. Wilson, *The Imperial Experience in Sub-Saharan Africa since 1870, Europe and the World in the Age of Expansion,* vols. 6, 7, and 8, ed. Boyd C. Shafer (Minneapolis: University of Minnesota Press, 1975, 1976, 1977). Also quite useful are L. H. Gann and Peter Duignan, *The Rulers of German Africa, 1884–1914* (Stanford: Stanford University Press, 1977); Mary Evelyn Townsend, *The Rise and Fall of Germany's Colonial Empire, 1884–1918* New York: Macmillan, 1930; rpt. New York: Howard Fertig, 1966); and Prosser Gifford and William R. Louis, eds., *Britain and Germany in Africa: Imperial Rivalry and Colonial Rule* (New Haven: Yale University Press, 1967).
21. Betts, *The False Dawn,* pp. 95–7, 106–14; Dodge, *Islands and Empires,* pp. 174–9.
22. Townsend, *Rise and Fall,* pp. 37–41; Alfred Funke, *Deutsche Siedlung über See: Ein Abriss ihre Geschichte und ihr Gedeihen in Rio Grande do Sul* (Halle:

Gebauer-Schwetschke, 1902); Karl Heinrich Oberacker, *Der deutsche Beitrag zum Aufbau der Brasilianischen Nation* (São Paulo: Herder Editora Livraria, 1955), esp. pp. 239–47; Karl A. Wettstein, *Brasilien und die deutsche-brasilienische Kolonie Blumenau* (Leipzig: Engelmann, 1907).

23. Gann and Duignan, *Rulers of German Africa*, pp. 26–7.
24. Ibid., pp. 45–55. See also Harry R. Rudin, *Germans in the Cameroons, 1884–1914: A Case Study of Modern Imperialism* (New Haven: Yale University Press, 1938), and Helmut Bley, *South-West Africa under German Rule, 1894–1914*, trans. Hugh Ridley (Evanston, IL: Northwestern University Press, 1971).
25. Kurt Hassert, *Deutschlands Kolonien: Erwebungs- und Entwickelungsgeschichte, Landes- und Volkskunde und wirtschaftliche Bedeutung unserer Schutzgebiete* (Leipzig: Seele, 1899), p. 296.
26. Gann and Duignan, *Rulers of German Africa*, pp. 27–9, 37.
27. Hartkopf, *Die Akamemie der Wissenschaften der DDR*, pp. 55–8; Grau, *Die Berliner Akademie der Wissenschaften in der Zeit der Imperialismus*, Ch. 1.
28. H. L. Zeijlstra, *Melchior Treub: Pioneer of a New Era in the History of the Malay Archipelago* (Amsterdam: Koninklijk Instituut voor de Tropen, 1959), pp. 11–32.
29. H. Graf zu Solms-Laubach, "Der Botanische Garten zu Buitenzorg auf Java," *Botanische Zeitung* 42 (1884):753–61, 769–80, 785–91.
30. Quoted in Zeijlstra, *Melchior Treub*, p. 54.
31. Ibid., pp. 123–6.
32. Ibid., pp. 55–6.
33. Ibid., pp. 123–6.
34. Haberlandt, "Anatomisch-physiologische Untersuchungen über das tropische Laubblatt, I. Ueber die Transpiration einiger Tropenpflanzen," *Sitzungsberichte der Akademie der Wissenschaften zu Wien*, math.-naturw. Classe, 101 (1892), 1:785–816; "II. Ueber wassersecernirende und- absorbirende Organe," ibid., 103 (1894), 1:489–538, 104 (1895), 1:55–116.
35. Haberlandt, *Eine botanische Tropenreise: Indo-malayische Vegetationsbilder und Reiseskizzen* (Leipzig: Engelmann, 1893), p. 1.
36. Ibid., p. 2.
37. Karl Goebel to Julius Sachs, September 28, 1886, in *Karl von Goebel: Ein deutsches Forscherleben in Briefen aus sechs Jahrzehnten, 1870–1932*, ed. Ernst Bergdolt (Berlin: Ahneuerbe-Stiftung Verlag, 1940), p. 63

6. Beyond Schwendener's circle: Ernst Stahl

1. M. Reess, "Anton de Bary," *Berichte der Deutschen Botanischen Gesellschaft* 6 (1888):viii–xxvi.
2. Hans Kniep, "Ernst Stahl," *Berichte der Deutschen Botanischen Gesellschaft* 37 (1919):(85)–(104); Wilhelm Detmer, "Ernst Stahl, seine Bedeutung als Botaniker und seine Stellung zu einigen Grundproblemen der Biologie," *Flora* 111–12 (1918):1–47; Karl Goebel, "Ernst Stahl zum Gedächtnis," *Die Naturwissenschaften* 8 (1920):141–6; and Otto Renner, "150 Jahre Botanische Anstalt zu Jena," *Jenaische Zeitschrift für Naturwissenschaft* 78 (1947):153–6.
3. For a brief historical sketch of the new University of Strasbourg, along with a list of the appropriations for the various departments, see S. Hausmann, "Die Kaiser Wilhelms-Universität zu Strassburg," in Lexis, ed., *Das Unterrichtswesen im Deutschen Reich* (Berlin: A. Asher, 1904), 1:599–606.
4. Reported by Hans Kniep, "Ernst Stahl," p. (101), and thereafter by nearly everyone else who wrote about Stahl.

5. Ernst Stahl, "Entwicklungsgeschichte und Anatomie der Lenticellen," *Botanische Zeitung* 31 (1873):561–8, 577–85, 593–601, 609–17.
6. Stahl, *Beiträge zur Entwicklungsgeschichte der Flechten*, 2 vols. (Leipzig: Arthur Felix, 1877).
7. In addition to defending his *Habilitationsschrift*, a paper on parasitic plants, Stahl had to field questions on a formidable list of botanical topics. This verbal part of his examination was all the more difficult for Stahl because German was his second language (he had spoken mainly French before the annexation of Alsace). To make matters worse, he had to endure this ordeal in public, in full academic dress, on a torrid summer afternoon. Vines, "Reminiscences of German Botanical Laboratories in the 'Seventies and 'Eighties of the Last Century," *The New Phytologist* 24 (1925):4–5; Kniep, "Ernst Stahl," p. (87).
8. See Julius Sachs, ed., *Arbeiten des Botanischen Instituts in Würzburg* (Leipzig: Engelmann, 1874–88), vols. 1 and 2.
9. A. C. Seward and F. F. Blackman, "Francis Darwin – 1848–1925," *Proceedings of the Royal Society* 110B (1932):i–xxi; Charles Darwin, assisted by Francis Darwin, *The Power of Movement in Plants* (New York: Appleton, 1881), pp. 418–92. Darwin's suggestion in this work that the phototropism of a young shoot depends upon a substance in the shoot tip that is sensitive to light and then transmits its effects downward (pp. 468–89) became the basis for the investigations into plant growth substances that so preoccupied plant physiologists during the first half of the twentieth century. See P. R. Bell, "The Movement of Plants in Response to Light," in *Darwin's Biological Work, Some Aspects Reconsidered* (Cambridge: Cambridge University Press, 1959), pp. 1–49.
10. Stahl, "Ueber den Einfluss des Lichtes auf die Bewegungserscheinungen der Schwärmsporen," *Verhandlungen der Physikalisch-Medizinische Gesellschaft in Würzburg* 12 (1878):269–70; idem, "Ueber den Einfluss des Lichtes auf die Bewegungen der Desmidien nebst einigen Bemerkungen über den richtenden Einfluss des Lichtes auf Schwärmsporen," ibid., 14 (1879):25–34; idem, "Ueber den Einfluss von Richtung und Stärke der Beleuchtung auf einige Bewegungserscheinungen im Pflanzenreiche," *Botanische Zeitung* 38 (1880):297–304, 321–43, 345–57, 361–8, 377–81, 393–400, 409–13.
11. Stahl, "Ueber den Einfluss der Lichtintensität auf Structur und Anordnung des Assimilationsparenchyms," *Botanische Zeitung* 38 (1880):868–74.
12. Goebel to de Bary, January 22, 1881, in *Karl von Goebel*, p. 25.
13. Wilhelm Lexis, *A General View of the History and Organization of Public Education in the German Empire*, trans. G. J. Tanson (Berlin: A. Asher, 1904), p. 45.
14. Roland Anheisser, *Natur und Kunst: Erinnerungen eines deutschen Malers* (Leipzig: Koehler and Ameling, 1937), p. 63. According to Anheisser's account, the Alsation students at Jena referred to Stahl as "Papa Stahl."
15. Ironically, when Stahl died in 1919, Goebel wrote to George Karsten: "Stahl's death has affected me quite deeply. He was my oldest friend, going back to my Strasbourg days (1876). Since the faculty is inquiring about a successor, I would gladly have placed my name on the list to escape this miserable Munich." Goebel added that he was too ill to leave Munich at that time. Goebel to Karsten, December 27, 1919, in *Karl von Goebel*, p. 176.
16. For a discussion of the place of Jena in the biological sciences in the nineteenth century, see George Uschmann, *Geschichte der Zoologie und zoologische Anstalten in Jena, 1779–1919* (Jena: G. Fischer, 1959), and Otto Renner, "150 Jahre Botanische Anstalt zu Jena."

17. Ernst Haeckel, *Generelle Morphologie der Organismen* (Berlin: Reimer, 1866), 2:286–7, 236n. See also Robert C. Stauffer, "Haeckel, Darwin, and Ecology," *Quarterly Review of Biology* 32 (1957):138–44.
18. Detmer, "Ernst Stahl," p. 11.
19. Stahl and Haberlandt met in Berlin in the summer of 1882, during the meeting that marked the founding of the Deutsche Botanische Gesellschaft. The two went on a long botanical excursion together and got along quite well. According to Haberlandt's account, Stahl, still under de Bary's influence, expressed surprise that they found each other so compatible. Haberlandt, *Erinnerungen: Bekenntnisse und Betrachtungen* (Berlin: Springer, 1933), p. 92.
20. Stahl, "Ueber den Einfluss des sonnigen oder schattigen Standortes auf die ausbildung der Laubblätter," *Jenaische Zeitschrift für Naturwissenschaft* 16 (1883):162–200.
21. Haberlandt, "Vergleichende Anatomie des Assimilatorischen Gewebesystems der Pflanzen," *Jahrbücher für wissenschaftliche Botanik* 13 (1881):80.
22. Detmer, "Ernst Stahl," p. 12.
23. Stahl, "Pflanzen und Schnecken: Eine biologische Studie über die Schutzmittel der Pflanzen gegen Schneckenfrass," *Jenaische Zeitschrift für Naturwissenschaft* 22 (1888):557–684; also published separately under the same title (Jena: G. Fischer, 1888).
24. Stahl, "Pflanzen und Schnecken," p. 557.
25. Ibid., p. 558.
26. Ibid., p. 670.
27. Ibid., p. 564.
28. Grisebach, *Die Vegetation der Erde nach ihrer klimatischen Anordnung: Ein Abriss der vergeleichenden Geographie der Pflanzen* (Leipzig: Engelmann, 1872), 1:443.
29. Stahl, "Pflanzen und Schnecken, pp. 565–6.
30. Ibid., p. 566.
31. Haberlandt, *Physiologische Pflanzenanatomie im Grundriss dargestellt* (Leipzig: Engelmann, 1884), p. 325.
32. Stahl, "Pflanzen und Schnecken," p. 684.
33. Goebel, "Ernst Stahl zum Gedächtnis," p. 143.
34. Stahl, "Regenfall und Blattgestalt," *Annales du jardin botanique de Buitenzorg* 11 (1893):98–183. A summary article under the same title appeared in *Botanische Zeitung* 51 (1893):145–52.
35. Stahl, "Regenfall und Blattgestalt," 120–1.
36. Ibid., pp. 121–8.
37. Ibid., p. 154.
38. Stahl, "Ueber bunte Laubblätter: Ein Beitrag zur Pflanzenbiologie, II," *Annales du jardin botanique de Buitenzorg* 13 (1896):137–216; idem, "Ueber den Pflanzenschlaf und verwandte Erscheinungen," *Botanische Zeitung* 55 (1897), 1:71–110.
39. Stahl, "Einige Versuche über Transpiration und Assimilation," *Botanische Zeitung* 52 (1894), 1:117–46.
40. Detmer, "Ernst Stahl," pp. 28–41.
41. Renner, "150 Jahre Botanische Anstalt zu Jena," p. 158.

7. Schimper and Schenck: from Bonn to Brazil

1. Heinrich Schenck, "A. F. Wilhelm Schimper," *Berichte der Deutschen Botanischen Gesellschaft* 19 (1903):(954)–(70); Schenck, "Wilhelm Schimper,"

Naturwissenschaftliche Rundschau 17 (1902):36–9; Percy Groom, "Prof. A. F. W. Schimper," *Nature* 64 (1901):551–2.

2. This was the problem first posed by Goethe and later worked out to reasonable satisfaction by Schwendener in *Mechanische Theorie der Blattstellungen.* For information concerning the work of the various Schimpers, see Karl Mägdefrau, *Geschichte der Botanik,* also *Dictionary of Scientific Biography* (New York: Scribner's, 1975), vol. 12.
3. A. F. W. Schimper, "Untersuchungen über die Entstehung der Stärkekörner," *Botanische Zeitung* 38 (1880):881–902, translated as "Researches upon the Development of Starch-Grains," *Quarterly Journal of Microscopical Science* 21 (1881):291–306.
4. Frederick O. Bower, "English and German Botany in the Middle and Towards the End of the Last Century," *The New Phytologist* 24 (1925):136.
5. Schenck, "A. F. Wilhelm Schimper," pp. (57)–(58). De Bary himself wrote the obituary for the elder Schimper in *Botanische Zeutung* 38 (1880):443–50.
6. C. R. Swanson, "A History of Biology at the Johns Hopkins University, *Bios* 22 (1951):248. The only evidence I have been able to find regarding Schimper's teaching duties at Johns Hopkins is a notice in *The Johns Hopkins University Circular,* no. 10 (April 1881), which lists him as offering a course in vegetable physiology.
7. George Karsten, "Eduard Strasburger," *Berichte der Deutschen Botanischen Gesellschaft* 30 (1912):(61)–(86); Hans Fitting, "Eduard Strasburger," in *Bonner Gelehrte: Beiträge zur Geschichte der Wissenschaften in Bonn, Mathematik und Naturwissenschaften. 150 Jahre Rheinische Friedrich-Wilhelms Universität zu Bonn, 1818–1968* (Bonn: Bouvier/Röhrscheid, 1970), pp. 246–57.
8. Strasburger, "Ueber die Bedeutung phylogenetischer Methoden für die Erforschung lebender Wesen," *Jenaische Zeitschrift für Naturwissenschaft* 8 (1874):56–80.
9. See William Coleman, "Cell Nucleus, and Inheritance: An Historical Study," Proceedings of the American Philosophical Society, 109 (1965): 145–54; also John Farley, *Gametes and Spores: Ideas About Sexual Reproduction, 1750–1914* (Baltimore: Johns Hopkins University Press, 1982), pp. 165–8.
10. Schenck, "A. F. Wilhelm Schimper," p. (59). Schenck did not give the name of the university.
11. Schimper, "Ueber Bau und Lebensweise der Epiphyten Westindiens," *Botanisches Centralblatt* 17 (1884):223–7, 253–8, 284–94, 318–26, 350–9.
12. Schimper, *Die epiphytische Vegetation Amerikas,* Botanische Mittheilungen aus den Tropen, vol. 2 (Jena: G. Fischer, 1888). I will be following Schimper's discussion of epiphytes in the 1888 work, esp. pp. 83–6 and 154–60. This discussion is restated in *Pflanzengeographie,* pp. 213–19 (*Plant Geography,* pp. 197–201). See note 41, this chapter, for the complete citation.
13. Schimper, *Plant Geography,* p. 199.
14. Ibid.
15. Ibid., p. 155.
16. August Weismann, "The Continuity of the Germ-Plasm as the Foundation of a Theory of Heredity" [1885] and "The Significance of Sexual Reproduction in the Theory of Natural Selection" [1886], in *Essays Upon Heredity and Kindred Problems,* trans. E. B. Poulton et al. (Oxford: Clarendon Press, 1889; rpt. Oceanside, NY: Dabor Science Publications, 1977); Frederick Churchill, "Rudolf Virchow and the Pathologist's Criteria for the Inheritance of Acquired Characteristics," *Journal of the History of Medicine* 31 (1976):117–48; Theodor Eimer,

Die Entstehung der Arten auf Grund von Vererbung erworbener Eigenschaften nach den Gesetzen organischen Wachsens (Jena: G. Fischer, 1888), translated as *Organic Evolution as the Result of the Inheritance of Acquired Characteristics according to the Laws of Organic Growth,* trans. James T. Cunningham (London: Macmillan, 1890).

17. Fritz Müller, *Für Darwin* (Leipzig: Engelmann, 1864). The English translation, *Facts and Arguments for Darwin,* trans. W. S. Dallas (London: John Murray, 1869), was financed by Darwin. Darwin also provided a preface for the translation of Herman Müller's work on insect pollination, *The Fertilisation of Flowers,* trans. and ed. D'Arcy W. Thompson (London: Macmillan, 1883).
18. Schimper, *Die Wechselbeziehungen zwischen Pflanzen und Ameisen im tropischen Amerika,* Botanische Mittheilungen aus den Tropen, vol. 1 (Jena: G. Fischer, 1888); Schenck, *Beiträge zur Biologie und Anatomie der Lianen im Besonderen der in Brasilien einheimischen Arten,* Botanische Mittheilungen aus den Tropen, vols. 4 and 5 (Jena: G. Fischer, 1892–3).
19. Alfred Möller, "Fritz Müller's Leben," in *Fritz Müller: Werke, Briefe, und Leben,* ed. A. Möller (Jena: G. Fischer, 1920), 3:130.
20. Schimper to Schenck, October 10, 1885, in Schenck, "A. F. Wilhelm Schimper," p. (61).
21. Schimper to D. C. Gilman, July 31, 1886, Daniel Coit Gilman Papers, Johns Hopkins University Archives.
22. M. Möbius, "Heinrich Schenck," *Berichte der Deutschen Botanischen Gesellschaft* 45 (1927):(89)–(101).
23. Schenck, *Die Biologie der Wassergewächse* (Bonn: Cohen, 1886); idem, *Vergleichende Anatomie der submersen Gewächse,* Biblioteca Botanica, vol. 1 (Cassel: Th. Fischer, 1886).
24. Wissenschaftliche Ergebnisse der Deutschen Tiefsee-Expedition auf dem Dampfer "Valdivia," 1898–99, vol. 2, ed. Carl Chun (Jena: G. Fischer, 1905–7); George Karsten and Heinrich Schenck, eds., *Vegetationsbilder* (Jena: G. Fischer, 1903–22); Schenck, "Die Gefässpflanzen der Deutschen Südpolar-Expedition, 1901–1903," in *Deutsche Südpolar-Expedition 1901–1903,* vol. 8, ed. Erich von Drygalski (Berlin: G. Reimer, 1906), pp. 97–123.
25. Schimper, *Die Wechselbeziehungen zwischen Pflanzen und Ameisen im tropischen Amerika.* A concise restatement of the main points of this work appeared in Schimper's *Pflanzengeographie,* pp. 154–68 (*Plant Geography,* pp. 140–53).
26. Schimper, *Plant Geography,* pp. 142–3.
27. Ibid., p. 144.
28. Ibid., p. 145.
29. Ibid.
30. William Morton Wheeler, *Ants: Their Structure, Development, and Behavior* (New York: Columbia University Press, 1910), pp. 299–310.
31. Schenck, *Beiträge zur Biologie und Anatomie der Lianen,* 2:3–14, 25–35. There is an excellent discussion of the anatomical peculiarities of lianas in Haberlandt, *Physiological Plant Anatomy,* 4th ed., trans. Montagu Drummond (London: Macmillan, 1914), pp. 690–6.
32. Müller, "Ueber das Holz einiger um Desterro wachsender Kletterpflanzen," *Botanische Zeitung* 24 (1866):57–60, 65–9. The anomalous secondary thickening of liana stems was first described in detail by de Bary in his *Vergleichende Anatomie der Vegetationsorgane der Phanerogamen und Farne* (Leipzig: Engelmann, 1877).
33. Schenck, *Beiträge zur Biologie und Anatomie der Lianen,* 2:1–39.

34. Ibid., 2:25.
35. Schenck, "A. F. Wilhelm Schimper," p. (64).
36. Schimper, "Ueber Schutzmittel des Laubes gegen Transpiration, besonders in der Flora Java's," *Sitzungsberichte der Akademie der Wissenschaften zu Berlin*, math.-phys. Klasse, 40 (1890):1045–62.
37. Ibid., pp. 1061–2.
38. Schimper, *Die indo-malayische Strandflora*, Botanische Mittheilungen aus den Tropen, vol. 3 (Jena: G. Fischer, 1891), p. 199.
39. Ibid., p. 201.
40. Schenck, "A. F. Wilhelm Schimper," pp. (64)–(68).
41. Schimper, *Pflanzengeographie auf physiologischer Grundlage* (Jena: G. Fischer, 1898); translated as *Plant Geography upon a Physiological Basis*, trans. W. R. Fisher, rev. and ed. Percy Groom and I. B. Balfour (Oxford: Clarendon Press, 1903). Although the title page states that the English edition is "revised and edited" by Groom and Balfour, the text is that of the German edition, as indicated in the preface to the English edition, p. ix.
42. Schimper, *Die epiphytische Vegetation Amerikas*, pp. 154–5.
43. Schimper, "Die gegenwärtigen Aufgaben der Pflanzengeographie," *Geographische Zeitschrift* 2 (1896):93.
44. Schimper, *Plant Geography*, p. v.
45. Ibid., p. vi.
46. Ibid., p. 781.

8. Teleology revisited? natural selection and plant adaptation

1. Karl Goebel, *Ueber Studium und Auffassung der Anpassungserscheinungen bei Pflanzen* (Munich: K. B. Akademie, G. Franz, 1898), p. 9.
2. Julius Wiesner, "The Relation of Plant Physiology to the Other Sciences," *Smithsonian Institution Annual Report* (1898), p. 432.
3. Immanuel Kant, *Critique of Judgement*, trans. J. H. Bernard (New York: Hafner, 1951), p. 258.
4. Timothy Lenoir, *The Strategy of Life: Teleology and Mechanics in Nineteenth Century German Biology* (Dordrecht and Boston: D. Reidel, 1982), esp. Chs. 1–4; see also Lenoir, "Teleology Without Regrets. The Transformation of Physiology in Germany: 1790–1847," *Studies in History and Philosophy of Science* 12 (1981):293–354.
5. Alexander von Humboldt, "Ideen zu einer Physiognomik der Gewächse," in *Ansichten der Natur mit wissenschaftlichen Erläuterungen* (Tübingen: J. G. Cotta, 1808; 2nd ed., 1826; 3rd ed., 1849). There were later editions as well, and many translations.
6. Grisebach first used the term *pflanzengeographische Formation* in "Ueber den Einfluss des Climas auf die Begränzung der natürlichen Floren," *Linnaea* 12 (1838):159–200.
7. Haberlandt, *Erinnerungen: Bekenntnisse und Betrachtungen* (Berlin: Springer, 1933), pp. 187–8; Anton Kerner, *Das Pflanzenleben der Donauländer* (Innsbruck: Wagner, 1863). Henry Conard, seeing in Kerner's work many of the fundamental ideas that later became part of ecological theory, translated the work as *The Background of Plant Ecology* (Ames: Iowa State University Press, 1951).
8. Kerner, *Background of Ecology*, p. 189.
9. Ibid., p. 35.

10. Grisebach, *Die Vegetation der Erde nach ihrer klimatischen Anordnung: Ein Abriss der vergeleichenden Geographie der Pflanzen* (Leipzig: Engelmann, 1872), 1:v.
11. Ibid., 1:368.
12. Ibid., 1:442.
13. Schimper, *Pflanzengeographie auf physiologischer Grundlage* (Jena: G. Fischer, 1898), p. iv.
14. Grisebach, *Die Vegetation der Erde,* 1:443.
15. Stahl, "Pflanzen und Schnecken: Eine biologische Studie über die Schutzmittel der Pfanzen gegen Schneckenfrass," *Jenaische Zeitschrift für Naturwissenschaft* 22 (1888):565.
16. Lenoir, *The Strategy of Life,* Ch. 4. See also Lenoir, "The Göttingen School and the Development of Transcendental Naturphilosophie in the Romantic Era," *Studies in History of Biology* 5 (1981):111–205; and William Coleman, "Bergmann's Rule: Animal Heat as a Biological Phenomenon," ibid., 3 (1979):67–88.
17. Grisebach, *Die Vegetation der Erde,* 1:iv.
18. Carl Nägeli, *Mechanische-physiologische Theorie der Abstammungslehre* (Munich: R. Oldenbourg, 1884).
19. August Weismann, "The Significance of Sexual Reproduction in the Theory of Natural Selection" [1886], trans. Selmar Schönland, in August Weismann, *Essays upon Heredity and Kindred Problems,* trans. E. B. Poulton et al. (Oxford: Clarendon Press, 1889; rpt. Oceanside, N.Y.: Dabor Science Publications, 1977), p. 259.
20. Julius Wiesner, "Die natürlichen Einrichtungen zum Schutze des Chlorophylls der Lebenden Pflanze," in *Festschrift zur Feier des fünfundzwanzigjährigen Bestehen der K. K. Zoologish-Botanischen Gesellschaft in Wien* (Vienna: W. Braunmüller, 1876), pp. 19–49; Anton Kerner, "Die Schutzmittel der Blüthen gegen unberufene Gäste," ibid., pp. 187–261. This latter work was translated as *Flowers and Their Unbidden Guests,* trans. W. Ogle (London: C. Kegan Paul, 1878).
21. Kerner, *Pflanzenleben,* 2 vols. (Leipzig: Verlag des bibliographischen Institutes, 1887–91); idem, *The Natural History of Plants: Their Forms, Growth, Reproduction, and Distribution,* trans. F. W. Oliver et al. (London: Blackie & Son, 1894–5).
22. This is a conclusion based upon my general reading of German biological literature from the late nineteenth century. Emmanuel Rádl, a strong critic of Darwinism, conveyed the same impression in *Geschichte der biologischen Theorien in der Neuzeit* (Leipzig: Engelmann, 1909; rpt. New York: Georg Olms, 1970), 2:406–13, 539–40.
23. Schimper, *Die epiphytische Vegetation Amerikas,* Botanische Mittheilungen aus den Tropen, vol. 2 (Jena: G. Fischer, 1888); see, for example, pp. 151–3.
24. Haberlandt, *Die Schutzeinrichtungen in der Entwickelung der Keimpflanze: Ein biologisch Studie* (Vienna: C. Gerold's Sons, 1877); Stahl, "Pflanzen und Schnecken;" Volkens, "Zur Kenntnis der Beziehungen zwischen Standort und anatomischen Bau der Vegetationsorgane," *Berlin Universität, Botanischer Garten, Jahrbuch* 3 (1884):1–46; Schimper, "Ueber Schutzmittel des Laubes gegen Transpiration besonders in der Flora Javas," *Sitzungsbereichte der Akademie der Wissenschaften zu Berlin,* phys.-math. Klasse 40 (1890):1045–62; Stahl, "Regenfall und Blattgestalt: Ein Beitrag zur Pflanzenbiologie," *Annales du jardin botanique de Buitenzorg* 11 (1893):98–183.

25. Karl Semper, *Animal Life as Affected by the Natural Conditions of Existence* (New York: D. Appleton, 1881), pp. 16–17. This is a translation of *Die natürlichen Existenzbedingungen der Thiere,* 2 vols. (Leipzig: Brockhaus, 1880), based on a series of lectures that Semper presented at the Lowell Institute in Boston in 1877.
26. Haberlandt, *Physiologische Pflanzenanatomie im Grundriss dargestellt* (Leipzig: Engelmann, 1984), p. 18.
27. Ibid., p. 24.
28. Lenoir argues that this sort of reasoning, involving a combination of energy physics and natural selection theory, led to the demise of the teleomechanist program in Germany in favor of a more thoroughly reductionist-mechanist approach under the leadership of Helmholtz and Du Bois-Reymond. *Strategy of Life,* Ch. 5.
29. Haberlandt, *Physiological Plant Anatomy,* p. 6.
30. Volkens, *Die Flora der ägyptisch-arabischen Wüste,* pp. 2–4.
31. Ibid., p. 2.
32. Schimper, *Die indo-malayische Strandflora,* Botanische Mittheilungen aus den Tropen, vol. 3 (Jena: G. Fischer), p. 201.
33. Tschirch, "Die Bedeutung der Blätter im Haushalte der Natur" [1893]. In *Vorträge und Reden von A. Tschirch gesammelt und herausgegeben von Schülern und Freunden* (Leipzig: Borntraeger, 1915), p. 55.
34. Haberlandt, "Vergleichende Anatomie des assimilatorischen Gewebesystems der Pflanzen." *Jahrbücher für wissenschaftliche Botank* 13 (1881):83.
35. Haberlandt, *Physiologische Pflanzenanatomie,* p. 17.
36. Haeckel, *Generelle Morphologie der Organismen: Allgemeine Grundzüge der organischen formen-Wissenschaft, mechanisch begründet durch die von Charles Darwin reformirte Descendenz-Theorie* (Berlin: Reimer, 1866), 1:64–88, 94–101; Sachs, *History of Botany* (1530–1860), tran. Henry E. F. Garnsy, rev. I. B. Balbour (Oxford: Clarendon Press, 1890), pp. 362–4ff. Sachs (p. 369) stated: "If the theory of descent finally liberated the morphological treatment of organisms from the influence of scholasticism, it is the theory of selection which has made it possible for physiology to set herself free from teleological explanations."
37. Haberlandt, *Physiologische Pflanzenanatomie,* p. 17.
38. Nägeli, *Mechanische-physiologische Theorie der Abstammungslehre,* p. 285.
39. Nägeli includes a lengthy discussion of the environmental alternatives, Darwinism and Lamarckism, on pp. 284–337.
40. Schimper, *Die epiphytische Vegetation Amerikas,* p. 155.
41. Stahl, "Regenfall und Blattgestalt," p. 154; "Pflanzen und Schnecken," p. 565.
42. See Peter J. Bowler, *The Eclipse of Darwinism: Anti-Darwinian Theories in the Decades around 1900* (Baltimore: Johns Hopkins University Press, 1983), pp. 58–106, and Rádl, *Geschichte der biologischen Theorien* 2:447–59.
43. Julius Sachs, "The General External Conditions of Plant Life," in *Lectures on the Physiology of Plants,* trans. Marshall Ward (Oxford: Clarendon Press, 1887), pp. 189–204.
44. Karl Goebel, *Pflanzenbiologische Schilderungen,* 2 vols. (Marburg: Elwert, 1889–91); idem, "Morphologische und biologische Studien," *Annales du jardin botanique de Buitenzorg* 7 (1887):1–40, and 9 (1891):1–126. Goebel's major work was *Organographie der Pflanzen, insbesondere der Archegoniaten und Samenpflanzen,* 2 vols. (Jena: G. Fischer, 1898–1901), *Organography of Plants, especially of the Archegoniatae and Spermatophyta,* trans. I. B. Balfour, 2 vols.

(Oxford: Clarendon Press, 1900–5). See George Karsten, "Karl Goebel," *Berichte der Deutschen Botanischen Gesellschaft* 50 (1932):(131–162).

45. Sachs, *Lectures on the Physiology of Plants,* p. 516.
46. Sachs, "Mechanomorphosen und Phylogenie (Ein Beitrag zur physiologischen Morphologie)," *Flora* 78 (1894):216.
47. Sachs developed these views in "Mechanomorphosen und Phylogenie" and "Phylogenetische Aphorismen und über innere Gestaltungsursachen oder Automorphosen," *Flora* 82 (1896):173–223. These two articles were part of a series, under the general title "Physiologische Notizen," that ran from 1892 to 1896 in various issues of *Flora.* After Sachs's death, Karl Goebel published the collection of articles as *Physiologische Notizen* (Marburg: Elwert, 1898).
48. Sachs, "Bemerkungen zum Anpassungsproblem" and "Bemerkungen zur Abstammungslehre," in Pringsheim, *Julius Sachs: Der Begründer der neueren Pflanzenphysiologie, 1832–1897* (Jena: G. Fischer, 1932), pp. 150–179.
49. Sachs, "Bemerkungen zum Anpassungsproblem," p. 156.
50. Sachs, "Bemerkungen zur Abstammungslehre," p. 175.
51. Ibid.
52. Goebel, "On the Study of Adaptation in Plants," *Science Progress* 1 (1894):189.
53. Haberlandt, *Physiologische Pflanzenanatomie,* p. 23.
54. Goebel, "On the study of Adaptation," p. 189.
55. Ibid.
56. Goebel, *Ueber Studium und Affassung der Anpassungserscheinungen bei Pflanzen* (Munich: K. B. Akademie, G. Franz, 1898), pp. 18–19.
57. Ibid., pp. 10–11.
58. Theordor Eimer, *Die Entstehung der Arten auf Grund von Vererbung erworbener Eigenschaften nach den Gesetzen organischen Wachsens* (Jena: G. Fischer, 1888), translated as *Organic Evolution as the Result of the Inheritance of Acquired Characters According to the Laws of Organic Growth,* trans. J. T. Cunningham (London: Macmillan, 1890).
59. Gustav Wolff, *Der gegenwärtige Stand des Darwinismus* (Leipzig: Engelmann, 1896); idem, *Beiträge zur Kritik der Darwin'schen Lehre: Gesammelte und vermehrte Abhandlung* (Leipzig: A. Geogi 1898); Ludwig Plate, *Ueber Bedeutung und Tragweite des Darwin'schen Selectionsprincips* (Leipzig: Engelmann, 1900); idem, *Selectionsprinzip und Probleme der Artbildung: Ein Handbuch der Darwinismus* (Leipzig: Engelmann, 1908); Richard von Wettstein, *Der Neo-Lamarckismus und seine Beziehungen zum Darwinismus* (Jena: G. Fischer, 1903); Adolf Wagner, *Geschichte des Lamarckismus, als Einführung in die psychobiologische Bewegung der Gegenwart* (Stuttgart: Franckh, 1908). Peter Bowler has provided an admirable discussion of neo-Lamarckism, orthogenesis, mutation theory, and other alternatives to natural selection in *The Eclipse of Darwinism.* Still useful for a discussion of the general state of evolution theory at the turn of the century is Vernon L. Kellogg, *Darwinism Today* (New York: Henry Holt, 1907).

9. The colonial connection: imperialism and plant adaptation

1. Lewis Pyenson has argued that scientific activity in colonial regions followed European practices more closely than economic and administrative activity. Pyenson, "Cultural Imperialism and Exact Sciences: German Expansion Overseas 1900–1930," *History of Science* 20 (1982):32–3.

2. Zeijlstra, *Melchior Treub: Pioneer of a New Era in the History of the Malay Archipelago* (Amsterdam: Koninklijk Instituut voor de Tropen, 1959), p. 59.
3. Schimper, "Die gegenwärtigen Aufgaben der Pflanzengeographie," *Geographische Zeitschrift* 2 (1896):90; idem, *Pflanzengeographie auf physiologischer Grundlage* (Jena: G. Fischer, 1898), p. iv; Tschirch, "Der javanische Urwald" [1890], in *Vorträge und Reden von A. Tschirch gesammelt und herausgegeben von Schülern und Freunden* (Leipzig: Borntraeger, 1915), p. 223. Haberlandt, *Eine botanische Tropenreise: Indo-malayische Vegetationsbilder und Reiseskizzen* (Leipzig: Engelmann, 1893), p. 3.
4. Zeijlstra, *Melchior Treub,* pp. 71–106.
5. Albrecht Zimmermann, "Das Kaiserliche Biologisch-Landwirtschaftliche Institut Amani," *Berichte der Deutschen Botanischen Gesellschaft* 22 (1904):532–6; Adolf Engler, "Das biologisch-landwirtschaftliche Institut zu Amani in Ost-Usambara," *Notizblatt des Königlichen botanischen Gartens und Museums zu Berlin* 4 (1903):63–6. Zimmermann's numerous research reports appear in various issues of *Der Pflanzer: Zeitschrift für Land- und Forstwirtschaft in Deutsch-Ost-Afrika* and in the Institute's *Mitteilungen,* both of which began publication in the first decade of the twentieth century. See also the discussion of the Amani Institute in L. H. Gann and Peter Duignan, *The Rulers of German Africa, 1884–1914* (Stanford: Stanford University Press, 1977), p. 189.
6. Georg Volkens, "Die Botanische Zentralstelle für did Kolonien, ihre Zweck und Ziele, *Jahresbericht der Vereinigung für angewandte Botanik* 5 (1907):32–48; Adolf Engler, "Die Botanische Zentralstelle für die deutschen Kolonien am kg. botan. Garten der Universität Berlin und die Entwicklung botanischer Versuchsstation in den Kolonien," *Botanische Jahrbücher* 15 (1893), Beiblatt 35:10–14; idem, "Bericht über die Tätigkeit der Botanischen Centralstelle für die Kolonien im Jahre 1901," *Notizblatt des Königlichen botanischen Gartens und Museums zu Berlin* 3 (1902):176–81; idem, "Bericht . . . 1902," ibid., 4 (1903): 215–24.
7. Eduard Strasburger, "The Development of Botany in Germany during the Nineteenth Century," trans. George J. Peirce, *Botanical Gazette* 20 (1895):257.
8. For the role of Kew Gardens in British colonial development, see Lucille Brockway, *Science and Expansion: The Role of the British Royal Botanic Gardens* (New York: Academic Press, 1979).
9. Adolf Engler, "Die Aufgaben grosser botanischer Gärten und Museum mit besonderer Berücksichtigung des Botanischen Gartens und Museums zu Dahlem," in *Der Königliche Botanische Garten und das Königlich Botanische Museum zu Dahlem* (Berlin: Horn and Raasch, 1909), pp. 1–10; P. Graebner and K. Peters, "Die Freiland-Anlagen," ibid., pp. 31–74; M. Gürke, "Die Nutzpflanzen- und Kolonialabteilung," ibid., pp. 131–42. For a description of the botanical garden and museum at the old location at Schöneberg, see Henry Potonié, "Der königliche botanische Garten zu Berlin," *Naturwissenschaftliche Wochenschrift* 5 (1890):211–13, 221–7, and idem, "Die pflanzengeographische Anlage im Kgl. botanischen Garten zu Berlin," ibid., pp. 254–5, 261–6. Also quite helpful is Zepernick and Karlsson, *Berlins Botanischer Garten* (Berlin: Haude and Spener, 1979), pp. 89–103.
10. For a discussion of the various institutes and programs associated with German colonial development, see Gann and Duignan, *Rulers of German Africa,* Chs. 8 and 9; Harry R. Rudin, *Germans in the Cameroons, 1884–1914: A Case Study in Modern Imperialism* (New Haven: Yale University Press, 1938), pp. 164–75, 252–5; and Helmut Bley, *South-West Africa under German Rule, 1894–1914,*

trans. Hugh Ridley (Evanston, IL: Northwestern University Press, 1971), pp. 105–11.

11. Eduard Sachau, "Das Seminar für orientalische Sprache," in Max Lenz, *Geschichte der Königlichen Friedrich-Wilhelms Universität zu Berlin* (Halle: Buchhandlung des Waissenhauses, 1910), 3:239–47.
12. Otto Warburg, "Was bezweckt die Zeitschrift für tropische Landwirtschaft?" *Der Tropenpflanzer* 1 (1897):1–4. Warburg coedited *Der Tropenpfanzer* from 1897 to 1922. After World War I he became an influential figure in the Zionist movement. See "Otto Warburg," obituary notice, *Nature* 141 (1938):191.
13. Adolf Engler, *Die Pflanzenwelt Afrikas, insbesondere seiner tropischen Gebiete* 5 vols. (Leipzig: Engelmann, 1908–25). For a complete list of Engler's publications and a description of his botanical career, see L. Diels, "Zum Gedächtnis von Adolf Engler," *Botanischer Jahrbücher* 64 (1931):I–LVI.
14. Engler's many reports on colonial botany are to be found in the *Notizblatt des Königlichen botanischen Gartens und Museums zu Berlin,* which began publication in 1895.
15. Tschirch, "Der javanische Urwald" [1890], in *Vorträge und Reden,* p. 223.
16. Haberlandt, *Eine botanische Tropenreise,* p. 2.
17. Haberlandt, *Erinnerungen: Bekenntnisse und Betrachtungen* (Berlin: Springer, 1933), p. 137.
18. Stahl, "Regenfall und Blattgestalt: Ein Beitrag zur Pflanzenbiologie," *Annales du jardin botanique de Buitenzorg* 11 (1893):98.
19. Haberlandt, *Eine botanische Tropenreise,* p. 3.
20. Ibid.
21. For a presentation of this argument by the contemporary German botanist whose work comes closest to that of Haberlandt, Schimper, and their colleagues, see Heinrich Walter, *Ecology of Tropical and Subtropical Vegetation,* trans. D. Mueller-Dombois, ed. J. H. Bernett (New York: Van Nostrand Reinhold, 1971), p. 72.
22. Schimper, *Die indo-malayische Strandflora,* Botanische Mitteilungen aus den Tropen, vol. 3 (Jena: G. Fischer, 1891), p. 6.
23. Schimper "Ueber Schutzmittel des Laubes gegen Transpiration besonders in der Flora Javas." *Sitzungsberichte der Akademie der Wisschschaften zu Berlin,* phys.-math. Klasse 40 (1890):1059–61.
24. Schimper, *Plant Geography Upon a Physiological Basis,* trans. W. R. Fisher, ed. and rev. P. Groom and I. B. Balfour (Oxford: Clarendon Press, 1903), p. vi.
25. Schenck, *Beiträge zur Biologie und Anatomie der Lianen, im besonderen der in Brasilien einheimischen Arten,* 2 vols., Botanische Mitteilungen aus den Tropen (Jena: G. Fischer, 1892–3), 1:3–4.
26. Tschirch, "Der javanische Urwald," p. 232. It is difficult to read these lines from Tschirch without thinking of the many corpses of native Africans and German soldiers sacrificed in skirmishes over control of East Africa, South-West Africa, Cameroon, and Togo.
27. See, for example, John Lussenhop, "Victor Hensen and the Development of Sampling Methods in Ecology," *Journal of the History of Biology* 7 (1974):319–37.

10. Toward a science of plant ecology

1. Percy Groom, "A. F. W. Schimper: An Appreciation," in Schimper, *Plant Geography,* p. xii.

2. The history of plant ecology has received considerable attention over the last decade. See Donald Worster, *Nature's Economy: The Roots of Ecology* (Cambridge: Cambridge University Press, 1985), Chs. 10 and 11; Ronald C. Tobey, *Saving the Prairies: The Life Cycle of the Founding School of American Plant Ecology, 1895–1955* (Berkeley: University of California Press, 1981); Malcolm Nicolson, "The Development of Plant Ecology, 1790–1960," (Ph.D. diss., University of Edinburgh, 1983); and Robert P. McIntosh, *The Background of Ecology: Concept and Theory* (Cambridge: Cambridge University Press, 1985). For sources on the history of ecology in general, see Frank N. Egerton, "The History of Ecology: Achievements and Opportunities, Part One," *Journal of the History of Biology* 16 (1983):259–310; and "Part Two," ibid. 18 (1985):103–43.
3. Carl Schröter, "Vorschläge der Berichterstatter," in *Phytogeographische Nomenklatur: Berichte und Vorschläge,* ed. Charles Flahault and Carl Schröter, IIIe Congrès International de Botanique, Bruxelles (Zürich: Zürche and Furrer, 1910), p. 24.
4. Arthur G. Tansley, Preface, *Types of British Vegetation,* ed. A. G. Tansley (Cambridge: Cambridge University Press, 1911), p. 3.
5. Some recent attention has been given to parallel developments in Russia that became known to the West only later in the twentieth century. See W. Carter Johnson and Norman R. French, "Soviet Union," in *Handbook of Contemporary Developments in World Ecology,* ed. Edward J. Kormondy and J. Frank McCormick (Westport, CT: Greenwood Press, 1981), pp. 343–83; V. D. Aleksandrova, "Russian Approaches to Classification of Vegetation," in *Classification of Plant Communities,* 2nd ed., ed. R. H. Whittaker (The Hague: W. Junk, 1978), pp. 169–200; and Douglas R. Weiner, *Models of Nature: Ecology, Conservation, and Cultural Revolution in Soviet Russia* (Bloomington: Indiana University Press, 1988), esp. Chs. 5 and 6.
6. For background on the various approaches to the classification and study of plant communities, the definitive work is Robert H. Whittaker, "Classification of Natural Communities," *Botanical Review* 28 (1962):1–239. See also Whittaker's much abbreviated version of this article, "Approaches to Classifying Vegetation," in Whittaker, *Classification of Plant Communities,* pp. 3–31. The latter volume and its companion, *Ordination of Plant Communities* (The Hague: W. Junk, 1978), together contain twenty articles on the various schools and approaches to plant ecology and plant sociology, most of which include historical introductions. Also quite useful is Nicolson, "Development of Plant Ecology," pp. 74–142. For turn-of-the-century views on these issues, see Robert Smith, "On the Study of Plant Associations," *Natural Science* 14 (1899):109–20; Charles Flahault, "A Project for Phytogeographic Nomenclature," *Bulletin of the Torrey Botanical Club* 24 (1901):157–92; and Charles E. Moss, "The Fundamental Units of Vegetation: Historical Development of the Concepts of the Plant Association and the Plant Formation," *New Phytologist* 9 (1910):18–53.
7. Eugenius Warming, *Plantesamfund: grundtrak of den ökologiska plantegeografi* (Copenhagen: Philipsens Forlag, 1895); idem, *Lehrbuch der ökologischen Pflanzengeographie: Eine Einführung in die Kenntnis der Pflanzenvereine,* trans. E. Knoblauch (Berlin: Borntraeger, 1896).
8. Schimper, "Bericht über die Fortschritte der Pflanzengeographie in den Jahren 1896 bis 1898," *Geographische Zeitschrift* 6 (1900):319.
9. R. G. Goodland of the New York Botanical Garden has made the claim that in his 1898 textbook Schimper borrowed heavily from Warming's works and used Warming's illustrations without proper acknowledgment. See Goodland, "The

Tropical Origin of Ecology: Eugen Warming's Jubilee," *Oikos* 26(1975):243. Goodland argues that "Warming's pioneer work in ecology has been obscured partly" by its inaccessibility and by Schimper's "hasty" publication; but, in fact, Warming remains the most widely and consistently acknowledged founder of the science of plant ecology. Goodland, incidentally, saw Warming's approach to plant ecology as having combined studies of environmental influences with Haberlandt's physiological-morphological treatment of adaptations (p. 244).

10. See William Coleman, "Evolution into Ecology? The Strategy of Warming's Ecological Plant Geography," *Journal of the History of Biology* 19 (1986):181–96. In my opinion, Ronald Tobey, *Saving the Prairies,* pp. 102–7, distorts the Darwinian emphasis in Warming's work in order to reinforce the dichotomy he tries to create between idealistic and mechanistic approaches to plant geography. His treatment of Warming as the central European proponent of the mechanistic school leads him to cast Warming as a strict Darwinian, which he was not. See D. Müller, "Eugenius Warming," *Dictionary of Scientific Biography* (New York: Scribner's, 1975), 14:181–2.
11. Franz Wilhelm Neger, *Biologie der Pflanzen auf experimenteller Grundlage (Bionomie)* (Stuttgart: F. Enke, 1913); Oscar Drude, "Franz Wilhelm Neger," *Berichte der Deutschen Botanischen Gesellschaft* 41 (1923):(84)–(92).
12. Hans Fitting, *Aufgabe und Ziele einer Vergleichenden Physiologie auf geographischer Grundlage* (Jena: G. Fischer, 1922); idem, *Die ökologische Morphologie der Pflanzen im Lichte neuerer physiologischer und pflanzengeographischer Forschungen* (Jena: G. Fischer, 1926).
13. Heinrich Walter, *Die Vegetation der Erde in ökologischer Betrachtung,* 2 vols. (Stuttgart: G. Fischer 1962–8). The first volume has been translated as *Ecology of Tropical and Subtropical Vegetation,* trans. D. Mueller-Dombois, ed. J. H. Burnett (New York: Van Nostrand Reinhold, 1971). See also Walter, *Vegetation of the Earth in Relation to Climate and the Eco-Physiological Conditions,* trans. Joy Wieser (New York: Springer-Verlag, 1973).
14. Walter, *Das Xerophytenproblem in Kausal-physiologischer Betrachtung* (Freising-Munich: F. P. Datterer, 1925); idem, *Einführung in die allgemeine Pflanzengeographie Deutschlands* (Jena: G. Fischer, 1927). Walter has recently published his memoirs, *Bekenntnisse eines Oekologen: Erlebtes in acht Jahrzehnten und auf Forschungsreisen in allen Erdteilen* (Stuttgart: G. Fischer, 1981). See pp. 41–83 regarding his early training and research in plant ecology. Walter's wife, Erna, also a botanist, was the daughter of Heinrich Schenck. From 1929 to 1932 Walter was a Rockefeller Fellow in the United States, where he worked with several American plant ecologists, including Frederic Clements.
15. Gottfried Reinhold Treviranus, *Biologie, oder Philosophie der lebenden Natur* (Göttingen: J. F. Röwer, 1802), 1:4; Jean-Baptiste Lamarck, *Hydrogéologie* (Paris: Published by the author, 1802), p. 8. See also William Coleman, *Biology in the Nineteenth Century: Problems of Form, Function, and Transformation* (Cambridge: Cambridge University Press, 1977), pp. 1–3.
16. In *Pflanzengeographie auf physiologischer Grundlage* (Jena: G. Fischer, 1898), p. IVn, Schimper explained that he was adopting *Oekologie* to replace *Biologie* as the term for the study of adaptations; and Haberlandt replaced the adjective *biologisch* with *ökologisch* in the second edition of *Eine botanische Tropenreise* (Leipzig: Engelmann, 1910).
17. Haeckel, *Generelle Morphologie der Organismen: Allgemeine Grundzüge der organischen formen-Wissenschaft, mechanisch begründet durch die von Charles Darwin reformirte Descendenz-Theorie* (Berlin: Reimer, 1866), 2:236; *Proceed-*

ings of the Madison Botanical Congress, Madison, Wisconsin, August 23 and 24, 1893 (Madison, WI: Published by the generosity of the citizens of Madison, 1894), pp. 35–8.

18. Henry Chandler Cowles, "An Ecological Aspect of the Concept of Species," *American Naturalist* 42 (1908):265; Roscoe Pound and Frederic E. Clements, *The Phytogeography of Nebraska*, 2nd ed. (Lincoln, NE: The Seminar, 1900), p. 161; Tansley, *Types of British Vegetation*, p. 2. For further elaboration of this theme within the context of the early development of formal plant ecology in the United States, see Eugene Cittadino, "Ecology and the Professionalization of Botany in America, 1890–1905," *Studies in History of Biology* 4 (1980):171–98.
19. V. M. Spalding, "The Rise and Progress of Ecology," *Science* 17 (1903):201–10; William F. Ganong, "The Cardinal Principles of Ecology," ibid., 19 (1904):493–8. Ganong's principles were all related to the problem of adaptation.
20. Frederic E. Clements, *Plant Physiology and Ecology* (New York: Holt, 1907), p. 1.
21. Drude's work, particularly *Deutschlands Pflanzengeographie* (Stuttgart: Engelhorn, 1896) and *Handbuch der Pflanzengeographie* (Stuttgart: Engelhorn, 1890), had an undeniable influence on Pound and Clements as they developed their studies of the Nebraska vegetation. See Roscoe Pound, "The Plant-Geography of Germany," *American Naturalist* 30 (1896):465–8.
22. This view has been reinforced recently by Joel B. Hagen. See Hagen, "Organism and Environment: Frederic Clements's Vision of a Unified Physiological Ecology," in *The American Development of Biology*, ed. R. Rainger, K. R. Benson, and J. Maienschein ((Philadelphia: University of Pennsylvania Press, 1988), pp. 257–80.
23. H. C. Cowles, "The Ecological Relations of the Vegetation on the Sand Dunes of Lake Michigan," *Botanical Gazette* 27 (1899):95–117, 162–202, 281–308, 361–91; idem, "The Physiographic Ecology of Chicago and Vicinity; a Study of the Origin, Development, and Classification of Plant Societies, ibid., 31 (1901):73–108, 145–82; idem, "The Causes of Vegetation Cycles," ibid. 51 (1911):161–83.
24. C. C. Adams and G. D. Fuller, "Henry Chandler Cowles, Physiographic Plant Ecologist," *Annals of the Association of American Geographers* 30 (1940):39–43; William S. Cooper, "Henry Chandler Cowles," *Ecology* 16 (1935):281–3; J. Ronald Engel, *Sacred Sands: The Struggle for Community in the Indian Dunes* (Middletown, CT: Wesleyan University Press, 1983), pp. 137–68.
25. *University of Chicago Annual Register, 1897–1898* (Chicago: University of Chicago Press, 1898), p. 322.
26. It is interesting to note that among twentieth-century plant ecologists, only Paul Sears, a student of Cowles, clearly identified Haberlandt as one of the founders of his discipline. Paul B. Sears, *The Biology of the Living Landscape: An Introduction to Ecology* (London: Allen & Unwin, 1964), pp. 67–8.
27. J. M. Coulter, C. R. Barnes, and H. C. Cowles, *A Textbook of Botany for Colleges and Universities*, vol. 2, *Ecology* (New York: American Book Co., 1911).
28. Charles Elton, *Animal Ecology* (New York: Macmillan, 1927), p. 1.
29. Paul Farber's suggestive paper on this theme, "The Transformation of Natural History in the Nineteenth Century," *Journal of the History of Biology* 15 (Spring, 1982):145–52, deserves serious attention by historians of science. See also David E. Allen, *The Naturalist in Britain: A Social History*. (London: Allen Lane–Penguin Books, 1976), Ch. 9, and Philip D. Lowe, "Amateurs and Professionals:

The Institutional Emergence of British Plant Ecology," *Journal of the Society for the Bibliography of Natural History* 7 (1976):517–35.

30. A brief introduction to such literature should include Arthur G. Tansley, "The Use and Abuse of Vegetational Concepts and Terms," *Ecology* 16 (1935):284–307; H. A. Gleason, "The Individualistic Concept of the Plant Association," *Bulletin of the Torrey Botanical Club* 53 (1926):7–26; Frank E. Egler, "Vegetation as an Object of Study," *Philosophy of Science* 9 (1942):245–60; idem, "A Commentary on American Plant Ecology, Based on the Textbooks of 1947–1949," *Ecology* 32 (1951):673–94; Stanley A. Cain, "The Climax and Its Complexities," *American Midland Naturalist* 21 (1939):146–81; and J. L. Harper, "A Darwinian Approach to Plant Ecology," *Journal of Ecology* 55 (1967):247–70.
31. Two other factors that need to be explored in greater depth are the influence of the early conservation movements and the influence of scientific agriculture and forestry. Donald Worster treats both factors in *Nature's Economy*, Chs. 11–13, as does Ronald Tobey in *Saving the Prairies*, Chs. 1 and 7, and in his earlier article "Theoretical Science and Technology in American Ecology," *Technology and Culture* 17 (1976):718–28; but both of these authors take up these themes at a relatively late stage. The influence of scientific agriculture and the increasing awareness of nature conservation and environmentalism in the sciences in the last half of the nineteenth century have hardly been addressed.
32. See Cittadino, "Ecology and the Professionalization of Botany," pp. 179–81, 192–4; Allen, *The Naturalist in Britain*, pp. 181–2, 242; and Hagen, "Organism and Environment."
33. For the mature expression of Clements's views on the climax as an integral biological entity, see Clements, *Plant Succession: An Analysis of the Development of Vegetation* (Washington: Carnegie Institution, 1916) and *Plant Competition: An Analysis of Community Functions* (Washington: Carnegie Institution, 1929). Tansley first presented the concept of the ecosystem in "The Use and Abuse of Vegetational Concepts and Terms," pp. 299–303. See also Tansley, "British Ecology During the Past Quarter-Century: The Plant Community and the Ecosystem," *Journal of Ecology* 27 (1929):513–30. Clements's and Tansley's functional approaches to the study of vegetation may have had their roots in other sources in addition to physiological botany. Both ecologists had serious interest in the behavioral sciences, Clements in Spencerian sociology, Tansley in Freudian psychology. See Tobey, *Saving the Prairies*, pp. 83–5, and Allen, *The Naturalist in Britain*, p. 243. Although Tobey and Allen have pointed out these interests and implied a connection to ecology, that connection has hardly been explored, whether for Clements and Tansley or for any of the other early ecologists.
34. See Frederick V. Coville and Daniel T. MacDougal, *The Desert Botanical Laboratory of the Carnegie Institution* (Washington: Carnegie Institution, 1903); also Tobey, *Saving the Prairies*, pp. 78–9, and Robert P. McIntosh, "Pioneer Support for Ecology," *Bioscience* 33 (1983):107–12.
35. The Ecological Society of America was formed in 1915, although its journal, *Ecology*, did not begin publication until 1920. That journal was the metamorphosed form of *Plant World*, edited by D. T. MacDougal, director of Botanical Research for the Carnegie Institution. The British Ecological Society and the British *Journal of Ecology* predate their American counterparts. Both the society and the journal were inaugurated in 1913.
36. W. H. Pearsall, "The Development of Ecology in Britain," *Journal of Ecology* 52 (1964), suppl., p. 2. Arthur G. Tansley, the dominant force in early British plant

ecology, acknowledged after an excursion to America in 1913 that, regarding ecology, "there can be little doubt that [America's] pre-eminence in this branch of biology – one of the most promising of all modern developments – will be maintained." Tansley, "International Phytogeographical Excursion (I.P.E.) in America, 1913," *The New Phytologist* 13 (1914):333. For a fuller discussion of particular developments in Britain, see Tansley, "British Ecology During the Past Quarter-Century," and Lowe, "Amateurs and Professionals."

37. Weiner, *Models of Nature*, esp. Chs. 6, 8, and 11. See also Weiner, "Community Ecology in Stalin's Russia: 'Socialist' and 'Bourgeois' Science," *Isis* 75 (1984): 684–96. The movement against holistic community ecology was led by I. I. Prezent, the same Soviet political thinker who later provided T. D. Lysenko with the ideological underpinnings for his campaign against Mendelian genetics in Russia.
38. Thomas Söderqvist, *The Ecologists: From Merry Naturalists to Saviours of the Nation* (Stockholm: Almqvist & Wiksell, 1986), p. 115. See Chs. 1 and 2, and esp. pp. 87–115.
39. Haberlandt, *Physiologische Pflanzenanatomie im Grundriss dargestelt,* 4th ed. (Leipzig: Engelmann, 1909), p. 7.
40. Julian Huxley, *Evolution: The Modern Synthesis* (London: Allen & Unwin, 1942), pp. 22–8.
41. Goebel, *Ueber Studium und Affassung der Anpassungserscheinungen bei Pflanzen* (Munich: K. B. Akademie, G. Franz, 1898), pp. 10–11; E. von Dennert, *Vom Sterbelager des Darwinismus* (Stuttgart: M. Kielmann, 1903). See Bowler, *The Eclipse of Darwinism: Anti-Darwinian Theories in the Decades around 1900* (Baltimore: Johns Hopkins University Press, 1983), and Vernon L. Kellogg, *Darwinism To-Day* (New York: Holt, 1907).

SELECT BIBLIOGRAPHY

The following is a list of the principal published and unpublished sources consulted in the preparation of this volume. This list is not intended to be exhaustive; the major emphasis is on primary sources and biographical material.

Bary, Heinrich Anton de. *Vergleichende Anatomie der Vegetationsorgane der Phanerogamen und Farne*. Leipzig: Engelmann, 1877. *Comparative Anatomy of the Vegetative Organs of the Phanerogams and Ferns*. Translated by F. O. Bower and D. H. Scott. Oxford: Clarendon Press, 1884.

Bergdolt, Ernst, ed. *Karl von Goebel: Ein deutsches Forscherleben in Briefen aus sechs Jahrzehnten, 1870–1932*. Berlin: Ahnenerbe-Stiftung Verlag, 1940.

Bower, Frederick O. "English and German Botany in the Middle and Towards the End of the Last Century." *The New Phytologist* 24 (1925):129–37.

Bowler, Peter J. *The Eclipse of Darwinism: Anti-Darwinian Theories in the Decades around 1900*. Baltimore: Johns Hopkins University Press, 1983.

Bradbury, S. *The Evolution of the Microscope*. Oxford: Pergamon Press, 1979.

Brockway, Lucille H. *Science and Expansion: The Role of the British Royal Botanic Gardens*. New York: Academic Press, 1979.

Cittadino, Eugene. "Ecology and the Professionalization of Botany in America, 1890–1905." *Studies in History of Biology* 4 (1980):171–98.

Coleman, William. "Cell, Nucleus, and Inheritance: An Historical Study." *Proceedings of the American Philosophical Society* 109 (1965):124–58.

"Morphology Between Type Concept and Descent Theory." *Journal of the History of Medicine* 31 (1976):149–75.

Correns, C., and H. Morstatt. "Albrecht Zimmermann." *Berichte der Deutschen Botanischen Gesellschaft* 49 (1931):(220)–(243).

Detmer, Wilhelm. "Ernst Stahl, seine Bedeutung als Botaniker und seine Stellung zu einigen Grundproblemen der Biologie." *Flora* 111–12 (1918):1–47.

Diels, L. "Zum Gedächtnis von Adolf Engler." *Botanischer Jahrbücher* 64 (1931):I–LVI.

Engler, Adolf. "Die Botanische Zentralstelle für die deutschen Kolonien am kgl. botan. Garten der Universität Berlin und die Entwicklung botanischer Versuchsstation in der Kolonien." *Botanische Jahrbücher* 15 (1893), Beibl. 35:10–14.

Die Pflanzenwelt Afrikas insbesondere seiner tropische Gebiete: Grundzüge der Pflanzenverbreitung in afrika und die Charakterpflanzen Afrikas. 5 vols. Die Vegetation der Erde: Sammlung pflanzengeographischen Monographen, vol. 9. Edited by Adolf Engler and Oscar Drude. Leipzig: Engelmann, 1908–25.

Farley, John. *Gametes and Spores: Ideas About Sexual Reproduction, 1750–1914*. Baltimore: Johns Hopkins University Press, 1982.

Funke, Alfred. *Deutsche Siedlung über See: Ein Abriss ihre Geschichte und ihr Gedeihen in Rio Grande do Sul*. Halle: Gebauer-Schwetschke, 1902.

Gann, Lewis H., and Peter Duignan. *The Rulers of German Africa, 1884–1914*. Stanford: Stanford University Press, 1977.

Goebel, Karl. "On the Study of Adaptation in Plants." *Science Progress* 1 (1894): 176–90.

"Ueber Leben und Werk von Julius Sachs." *Flora* 84 (1897):101–30.

Ueber Studium und Auffassung der Anpassungserscheinungen bei Pflanzen. Munich: K. B. Akademie, G. Franz, 1898.

"Ernst Stahl zum Gedächtnis." *Die Naturwissenschaften* 8 (1920):141–6.

Wilhelm Hofmeister: The Work and Life of a Nineteenth Century Botanist. Translated by H. M. Bower. London: Ray Society, 1926.

Grau, Conrad. *Die Berliner Akademie der Wissenschaften in der Zeit des Imperialismus*. Berlin: Akademie-Verlag, 1975.

Green, J. Reynolds. *A History of Botany, 1860–1900: Being a Continuation of Sachs 'History of Botany, 1530–1860.'* Oxford: Clarendon Press, 1909.

Grisebach, August. "Ueber den Einfluss des Climas auf die Begränzung der natürlichen Flora." *Linnaea* 52 (1838):159–200.

Die Vegetation der Erde nach ihrer klimatischen Anordnung: Ein Abriss der vergleichenden Geographie der Pflanzen. 2 vols. Leipzig: Engelmann, 1872.

Guttenberg, Herman von. "Gottlieb Haberlandt," *Phyton* 6 (1955):1–14.

Haberlandt, Gottlieb. *Die Schutzeinrichtungen in der Entwickelung der Keimpflanze: Eine biologische Studie*. Vienna: C. Gerold's Sohn, 1877.

Die Entwickelungsgeschichte des mechanischen Gewebesystems der Pflanzen. Leipzig: Engelmann, 1879.

"Vergleichenden Anatomie des assimilatorischen Gewebesystems der Pflanzen." *Jahrbücher für wissenschaftliche Botanik* 13 (1881):74–188.

"Die physiologischen Leistungen der Pflanzengewebe." In *Handbuch der Botanik*, edited by A. Schenk, 2:558–693. The Schenk volumes appeared as part of the *Encyclopaedie der Wissenschaften*, edited by G. Jaeger et al. Breslau: E. Trewendt, 1882.

Physiologische Pflanzenanatomie im Grundriss dargestellt. Leipzig: Engelmann, 1884. 2nd ed., 1896; 3rd ed., 1904; 4th ed., 1909; 5th ed., 1918; 6th ed., 1924. *Physiological Plant Anatomy*. 4th ed. Translated by Montagu Drummond. London: Macmillan, 1914.

"Anatomisch-physiologische Untersuchungen über das tropische Laubblatt, I. Ueber die Transpiration einiger Tropenpflanzen," and "II. Ueber die wassersecernirende und- absorbirende Organe." *Sitzungsberichte der Akademie der Wissen schaften zu Wien,* math.-naturw. Classe 101 (1892), I:785–816; 103 (1894), I:489–538; and 104 (1895), I:55–116.

Eine botanische Tropenreise: Indo-malayische Vegetationsbilder und Reiseskizzen. Leipzig: Engelmann, 1893.

Erinnerungen: Bekenntnisse und Betrachtungen. Berlin: Springer, 1933.

Haeckel, Ernst. *Generelle Morphologie der Organismen: Allgemeine Grundzüge der organischen formen-Wissenschaft, mechanisch begründet durch die von Charles Darwin reformirte Descendenz-Theorie.* 2 vols. Berlin: Reimer, 1866.

Harms, H. "Nachschrift [for Georg Volkens]." *Verhandlungen des Botanischen Vereins der Provinz Brandenburg* 59 (1917):1–23. Includes a brief autobiographical sketch by Volkens.

Hartkopf, Werner. *Die Akademie der Wissenschaften der DDR: Ein Beitrag zu ihrer Geschichte.* Berlin: Akademie-Verlag, 1975.

Hassert, Kurt. *Deutschlands Kolonien: Erwerbungs- und Entwickelungsgeschichte, Landes-und Volkskunde und wirtschaftliche Bedeutung unserer Schutzgebiete.* Leipzig: Seele, 1899.

Heinricher, Emil. "Ueber isolateralen Blattbau mit besonderer Berücksichtigung der europäischen, speziell der deutschen Flora: Ein Beitrag zur Anatomie und Physiologie des Laubblätter." *Jahrbücher für wissenschaftliche Botanik* 15 (1884): 502–67.

"Zur Kenntnis des Blattbaues der Alpenpflanzen und dessen biologischer Bedeutung." *Sitzungsberichte der Akademie der Wissenschaften zu Wien,* math.-naturw. Classe, 101 (1892), 1:487–548.

Hofmeister, Wilhelm. *Vergleichende Untersuchungen der Keimung, Entfaltung und Fruchtbildung höherer Kryptogamen . . . und der Samenbildung der Coniferen.* Leipzig: Friedrich Hofmeister, 1851. *On the Germination, Development, and Fructification of the Higher Cryptogamia, and on the Fructification of the Coniferae.* 2nd ed. (not published in German). Translated by Frederick Currey. London: Ray Society, 1862.

Hughs, Arthur. "Studies in the History of Microscopy, I. The Influence of Achromatism" and "II. The Later History of the Achromatic Microscope." *Journal of the Royal Microscopal Society* 75 (1955):1–22; 76 (1955):47–60.

Humboldt, Alexander von. *Ansichten der Natur mit wissenschaftlichen Erläuterungen.* Tübingen: J. G. Cotta, 1808; 2nd ed., 1826; 3rd. ed., 1849. *Views of Nature, or Contemplations on the Sublime Phenomena of Creation, with Scientific Illustrations.* Translated by E. C. Otté. London: Henry G. Bohn, 1850.

and Aimé Bonpland. *Essai sur la géographie des plantes.* Paris: F. Schoell, 1807. Reprint. New York: Arno Press, 1977.

Karsten, George. "Eduard Strasburger." *Berichte der Deutschen Botanischen Gesellschaft* 30 (1912):(61)–(86).

Kelly, Alfred. *The Descent of Darwin: The Popularization of Darwinism in Germany, 1860–1914.* Chapell Hill: University of North Carolina Press, 1981.

Kerner, Anton Joseph, Ritter von Marilaun. *Das Pflanzenleben der Donauländer.* Innsbruck: Wagner, 1863. *The Background of Plant Ecology.* Translated by Henry Conard. Ames: Iowa State University Press, 1951.

"Die Schutzmittel der Blüthen gegen unberufene Gäste." In *Festschrift der K. K. Zoologisch-botanische Gesellschaft in Wien*, 187–261. Vienna: W. Branmüller, 1876. Published separately under the same title, Vienna: Zoologisch-botanische Gesellschaft, 1876. *Flowers and Their Unbidden Guests*. Translated, revised, and edited by W. Ogle. London: C. Kegan Paul, 1878.

Pflanzenleben. 2 vols. Leipzig: Verlag des bibliographischen Instituts, 1887–91. *The Natural History of Plants: Their Forms, Growth, Reproduction, and Distribution*. 2 vols. Translated by F. W. Oliver et al. London: Blackie & Son, 1894–5.

Kniep, Hans. "Ernst Stahl." *Berichte der Deutschen Botanischen Gesellschaft* 37 (1919):(85)–(104).

Der Königliche botanische Garten und das königliche botanische Museum zu Dahlem. Edited by the Ministerium der geistlichen, Unterrichts- und medizinal-Angelegenheiten. Berlin: Horn and Raasch, 1909.

Lenoir, Timothy. "Kant, Blumenbach, and Vital Materialism in German Biology." *Isis* 71 (1980):77–108.

"Teleology without Regrets. The Transformation of Physiology in Germany: 1790–1847." *Studies in History and Philosophy of Science* 12 (1981):293–354.

The Strategy of Life: Teleology and Mechanics in Nineteenth-Century German Biology. Dordrecht and Boston: D. Reidel, 1982.

Lexis, Wilhelm. *Die Deutschen Universitäten*. 2 vols. Berlin: A. Asher, 1893.

A General View of the History and Organization of Public Education in the German Empire. Translated by G. J. Tanson. Berlin: A. Asher, 1904.

ed. *Das Unterrichtswesen im Deutschen Reich, I. Die Universitäten im Deutschen Reich*. Berlin: A. Asher, 1904.

Mägdefrau, Karl. *Geschichte der Botanik: Leben und Leistung grosser Forscher*. Stuttgart: G. Fischer, 1973.

McClelland, Charles E. *State, Society, and University in Germany, 1700–1914*. Cambridge: Cambridge University Press, 1980.

Möbius, M. "Heinrich Schenck." *Berichte der Deutschen Botanischen Gesellschaft* 45 (1927):(89)–(101).

Möller, Alfred. "Fritz Müller's Leben." In *Fritz Müller: Werke, Briefe, und Leben*, edited by A. Möller. Jena: G. Fischer, 1892–3.

Montgomery, William M. "Germany." In *The Comparative Reception of Darwinism*, edited by Thomas F. Glick, 81–116. Austin: University of Texas Press, 1974.

"Evolution and Darwinism in German Biology, 1800–1883." Ph.D. diss., University of Texas, 1975.

Mullen, Pierce C. "The Preconditions and Reception of Darwinian Biology in Germany, 1800–1870." Ph.D. diss., University of California, 1964.

Müller, Fritz. *Für Darwin*. Leipzig: Engelmann, 1864. *Facts and Arguments for Darwin*. Translated by W. S. Dallas. London: John Murray, 1869.

Müller, Hermann. *Die Befruchtung der Blumen durch Insekten*. 8 vols. Leipzig: Engelmann, 1873. *The Fertilisation of Flowers*. Translated and edited by D'Arcy W. Thompson. London: Macmillan, 1883.

Nägeli, Carl. *Entstehung und Begriff der Naturhistorischen Art*. 2nd ed. Munich: Verlage der Königl. Akademie, 1865.

Mechanisch-physiologische Theorie der Abstammungslehre. Munich: R. Oldenbourg, 1884.

and Simon Schwendener. *Das Mikroskop*. 2 parts. Leipzig: Engelmann, 1865–7. 2nd ed. 1877.

Nicolson, Malcolm. "The Development of Plant Ecology, 1790–1960." Ph.D. diss., University of Edinburgh, 1983.

Paulsen, Friedrich. *The German Universities and University Study*. Translated by F. Thilly and W. W. Elwang. New York: Charles Scribner's Sons, 1906.

Potonié, Henry. "Der königliche botanische Garten zu Berlin." *Naturwissenschaftliche Wochenschrift* 5 (1980):211–13, 221–7.

Pringsheim, Ernst G. *Julius Sachs: Der Begründer der neueren Pflanzenphysiologie, 1832–1897*. Jena: G. Fischer, 1932.

Rádl, Emmanuel. *Geschichte der biologischen Theorien*. 2 vols. Leipzig: Engelmann, 1905–9. Reprint. New York: Georg Olms, 1970.

Reess, M. "Anton de Bary." *Berichte der Deutschen Botanischen Gesellschaft* 6 (1888):viii–xxvi.

Reinhardt, Otto. "Georg Volkens." *Berichte der Deutschen Botanischen Gesellschaft*. 35 (1917):(65)–(82).

Renner, Otto. "150 Jahre botanische Anstalt zu Jena." *Jenaische Zeitschrift für Naturwissenschaft* 78 (1947):131–62.

Ringer, Fritz K. *The Decline of the German Mandarins: The German Academic Community, 1890–1933*. Cambridge: Harvard University Press, 1969.

Rudin, Harry R. *Germans in the Cameroons, 1884–1914: A Case Study in Modern Imperialism*. New Haven: Yale University Press, 1938.

Sabalitschka, T. "Alexander Tschirch." *Berichte der Deutschen Botanischen Gesellscahft* 59 (1941–2):(67)–(108).

Sachs, Julius. *Lehrbuch der Botanik, nach dem gegenwärtigen Stand der Wissenschaft*. Leipzig: Engelmann, 1868. *Text-Book of Botany, Morphological and Physiological*. 4th ed. Translated by Alfred W. Bennet and W. T. Thiselton Dyer. Oxford: Clarendon Press, 1875.

Ueber den gegenwärtigen Zustand der Botanik in Deutschland. Würzburg: F. E. Thein, 1872.

Vorlesungen über Pflanzen-Physiologie. Leipzig: Engelmann, 1882. *Lectures on the Physiology of Plants*. Translated by Marshall Ward. Oxford: Clarendon Press, 1887.

Physiologische Notizen. Marburg: N. G. Elwert, 1898.

History of Botany (1530–1860). Translated by Henry E. F. Garnsy, revised by I. B. Balfour. Oxford: Clarendon Press, 1890.

"Bemerkungen zum Anpassungsproblem" and "Bemerkungen zur Abstammungslehre." In E. G. Pringsheim, *Julius Sachs*, 150–79.

Schenck, Heinrich. *Die Biologie der Wassergewächse*. Bonn: Cohen, 1886.

Vergeleichende Anatomie der submersen Gewächse. Bibliotheca Botanica, vol. 1. Cassel: Th. Fischer, 1886.

Beiträge zur Biologie und Anatomie der Lianen, im besonderen der in Brasilien einheimischen Arten. 2 vols. Botanische Mitteilungen aus den Tropen, vols. 4 and 5. Jena: G. Fischer, 1892–3.

"A. F. W. Schimper." *Berichte der Deutschen Botanischen Gesellschaft* 19 (1901):(954)–(970).

"Wilhelm Schimper." *Naturwissenschaftliche Rundschau* 17 (1902):36–9.

Schimper, Andreas Franz Wilhelm. "Untersuchungen über die Entstehung der Stärkekörner." *Botanische Zeitung* 38 (1880):881–902. Translated as "Researches upon the Development of Starch-Grains." *Quarterly Journal of Microscopical Science* 21 (1881):291–306.

"Ueber Bau und Lebensweise der Epiphyten Westindiens." *Botanisches Centralblatt* 17 (1884):223–7, 253–8, 284–94, 319–26, 350–9, 381–8.

Die epiphytische Vegetation Amerikas. Botanische Mitteilungen aus den Tropen, vol. 2. Jena: G. Fischer, 1888.

Wechselbeziehungen zwischen Pflanzen und Ameisen. Botanische Mitteilungen aus den Tropen, vol. 1. Jena: G. Fischer, 1888.

"Ueber Schutzmittel des Laubes gegen Transpiration besonders in der Flora Javas." *Sitzungsberichte der Akademie der Wissenschaften zu Berlin*, phys.-math. Klasse 40 (1890):1045–62.

Die indo-malayische Strandflora. Botanische Mitteilungen aus den Tropen, vol. 3. Jena: G. Fischer, 1891.

Pflanzengeographie auf physiologischer Grundlage. Jena: G. Fischer, 1898. *Plant Geography Upon a Physiological Basis*. Translated by W. R. Fisher, edited and revised by P. Groom and I. B. Balfour. Oxford: Clarendon Press, 1903.

Schleiden, Matthias Jacob. *Grundzüge der wissenschaflichen Botanik nebst einer methodologischen Einleitung als Anleitung zum Studium der Pflanze*. 2 vols. Leipzig: Engelmann, 1842–3. *Principles of Scientific Botany or Botany as an Inductive Science*. 2nd. ed. Translated by Edwin Lankester. London, 1849. Reprint. New York: Johnson Reprint Corp., 1969.

Schwendener, Simon. *Das mechanische Princip im anatomischen Bau der Monocotylen*. Leipzig: Engelmann, 1874.

Mechanische Theorie der Blattstellungen. Leipzig: Engelmann, 1878.

Gesammelte Botanische Mittheilungen. 2 vols. Berlin: Borntraeger, 1898.

Scott, D. H. "German Reminiscences of the Early 'Eighties." *The New Phytologist* 24 (1925):9–16.

Semper, Karl G. *Die natürlichen Existenzbedingungen der Thiere*. 2 vols. Leipzig: F. A. Brockhaus, 1880. *Animal Life as Affected by the Natural Conditions of Existence*. New York: Appleton, 1881.

Söderqvist, Thomas. *The Ecologists: From Merry Naturalists to Saviours of the Nation*. Stockholm: Almqvist and Wiksell, 1986.

Sperlich, Adolf. "Emil Heinricher." *Berichte der Deutschen Botanischen Gesellschaft* 52 (1934):(188)–(205).

Stahl, Ernst. "Ueber den Einfluss der Lichtintensität auf Structur und Anordnung des Assimilationsparenchyms." *Botanische Zeitung* 38 (1880):868–74.

"Ueber den Einfluss von Richtung und Stärke der Beleuchtung auf einige Bewegungserscheinungen im Pflanzenreiche." *Botanische Zeitung* 38 (1880):297–304, 321–43, 345–57, 361–8, 377–81, 393–400, 409–13.

Ueber den Einfluss des sonnigen und schattigen Standortes auf die Ausbildung der Laubblätter." *Jenaische Zeitschrift für Naturwissenschaften* 16 (1882):162–200.
"Pflanzen und Schnecken: Eine biologische Studie über die Schutzmittel der Pflanzen gegen Schneckenfrass." *Jenaische Zeitschrift für Naturwissenschaften* 22 (1888):557–684. Published separately under the same title, Jena: G. Fischer, 1888.
"Einige Versuche über Transpiration und Assimilation." *Botanische Zeitung* 52 (1894), I:117–46.
"Regenfall und Blattgestalt: Ein Beitrag zur Pflanzenbiologie." *Annales du jardin botanique de Buitenzorg* 11 (1893):98–183.
Stauffer, Robert C. "Haeckel, Darwin, and Ecology." *Quarterly Review of Biology* 32 (1957):138–44.
Strasburger, Eduard. "Ueber die Bedeutung phylogenetischer Methoden für die Erforschung lebender Wesen." *Jenaische Zeitschrift für Naturwissenschaft* 8 (1874):56–80.
"Botanik." In *Die Deutschen Universitäten,* edited by Wilhelm Lexis, II:73–94. "The Development of Botany in Germany in the Nineteenth Century." Translated by George J. Peirce. *Botanical Gazette* 20 (1895):193–204, 249–57.
Townsend, Mary E. *The Rise and Fall of Germany's Colonial Empire, 1884–1918.* New York: Macmillan, 1930. Reprint. New York: Howard Fertig, 1966.
Tshirch, Alexander. "Ueber einige Beziehungen des anatomischen Baues der Assimilationsorgane zu Klima und Standort, mit specieller Berücksichtigung des Spaltöffnungsapparates." *Linnaea* 43 (1880–2):139–252.
"Der javanische Urwald" [1890]. In *Vorträge und Reden,* 222–35.
"Physiologische Studien über die Samen, insbesonder die Saugorgane derselben." *Annales du jardin botanique de Buitenzorg* 9 (1891):143–83.
"Die Bedeutung der Blätter im Haushalte der Natur" [1893]. In *Vorträge und Reden,* pp. 52–82.
Vorträge und Reden von A. Tschirch gesammelt und herausgegeben von Schülern und Freunden. Leipzig: Borntraeger, 1915.
Erlebtes und Erstrebtes: Lebenserinnerungen. Bonn: Friedrich Cohn, 1921.
Uschmann, Georg. *Geschichte der Zoologie und zoologishen Anstalten in Jena, 1779–1919.* Jena: G. Fischer, 1959.
Vines, Sidney H. "Reminiscences of German Botanical Laboratories in the 'Seventies and 'Eighties of the Last Century." *The New Phytologist* 24 (1925):1–8.
Volkens, Georg. "Ueber Wasserausscheidung in liquider Form an den Blättern höherer Pflanzen." *Berlin Universität, Botanischer Garten, Jahrbuch* 2 (1883):166–209.
"Zur Kenntnis der Beziehungen zwischen Standort und anatomischen bau der Vegetationsorgane." *Berlin Universität, Botanischer Garten, Jahrbuch* 3 (1884):1–46.
"Zur Flora der ägyptisch-arabischen Wüste: Eine vorläufige Heinrich *Sitzungsberichte der Akademie der Wissenschaften zu Berlin,* math.-phys. Klasse 28 (1886):63–82.
Die Flora der ägyptisch-arabischen Wüste auf Grundlage anatomisch-physiologischer Forschungen dargestellt. Berlin: Gebrüder Borntraeger, 1887.

Der Kilimandscharo: Darstellung der allgemeineren Ergebnisse eines fünfzehnmonatigen Aufenthalts im Dschaggalande. Berlin: D. Reimer, 1897.
"Die Botanische Zentralstelle für die Kolonien, ihre Zwecke und Ziele." *Jahresbericht der Vereinigung für angewandte Botanik* 5 (1907):32–48.
Walter, Heinrich. *Die Vegetation der Erde in ökologischer Betrachtung*. 2 vols. Stuttgart: G. Fischer, 1962–8.
Bekenntnisse eines Oekologen: Erlebtes in acht Jahrzehnten und auf forschungsreisen in allen Erdteilen. Stuttgart: Fischer, 1981.
Warming, Eugenius. *Lehrbuch der ökologischen Pflanzengeographie: Eine Einführung in die Kenntnis der Pflanzenvereine*. Translated by E. Knoblauch. Berlin: Borntraeger, 1896. *Oecology of Plants: An Introduction to the Study of Plant-Communities*. Translated and edited by P. Groom and I. B. Balfour. Oxford: Clarendon Press, 1909.
Weiner, Douglas R. *Models of Nature: Ecology, Conservation, and Cultural Revolution in Soviet Russia*. Bloomington: Indiana University Press, 1988.
Weismann, August. *Essays upon Heredity and Kindred Biological Problems*. Translated by Edward B. Poulton et al. Oxford: Clarendon Press, 1889.
Zeijlstra, H. H. *Melchior Treub: Pioneer of a New Era in the History of the Malay Archipelago*. Amsterdam: Koninklijk Instituut voor de Tropen, 1959.
Zimmerman, Albrecht. "Ueber mechische Einrichtungen zur Verbreitung der Samen und Früchte mit besonderer Berücksichtigung der Torsionserscheinungen." *Jahrbücher für wissenschaftliche Botanik* 12 (1881):542–77.
"Das Kaiserliche Biologisch-Landwirtschaftliche Institut Amani." *Berichte der Deutschen Botanischen Gesellscahft* 22 (1904):532–6.
"Simon Schwendener." *Berichte der Deutschen Botanischen Gesellschaft* 40 (1922):(55)–(76).

INDEX